REED'S
Engineering Drawing for Marine Engineers

By **H. G. BECK**, C.Eng, F.I.Mar.E., M.R.I.N.A.

THOMAS REED PUBLICATIONS
A DIVISION OF THE ABR COMPANY LIMITED

In the same Series:			Vol. 6	**Basic Electrotechnology**
Vol. 1	**Mathematics**		Vol. 7	**Advanced Electrotechnology**
Vol. 2	**Applied Mechanics**		Vol. 8	**General Engineering Knowledge**
Vol. 3	**Applied Heat**		Vol. 9	**Steam Engineering Knowledge**
Vol. 4	**Naval Architecture**		Vol. 10	**Instrumentation and Control Systems**
Vol. 5	**Ship Construction**		Vol. 12	**Motor Engineering Knowledge**

First Edition 1973

Second Edition 1978

Reprinted 1996

Reprinted 2002

ISBN 0 901281 66 2

© Thomas Reed Publications
REED's is the trademark of The ABR Company Limited

THOMAS REED PUBLICATIONS
The Barn, Ford Farm, Bradford Leigh
Bradford-on-Avon, Wiltshire BA15 2RP
United Kingdom
E-mail: sales@abreed.demon.co.uk

Produced by Omega Profiles Ltd, SP11 7RW
Printed and Bound in Great Britain

ACKNOWLEDGEMENTS

The thanks of the publishers are due to The British Standards Institution for permission to reprint certain sections from BS308: Engineering Drawing Practice, and BS3692: ISO Metric Precision Hexagon Bolts, Screws and Nuts.

Acknowledgement is also made to the following firms who supplied drawings of some of the items included in this book:

Harland & Wolff Ltd **J. & W. Kirkham Ltd**

John Hastie & Co Ltd **Howden Group Ltd**

Vickers Ltd—Michell Bearings **The Welin Davit & Eng. Co Ltd**

J. & E. Hall Products **Doxford & Sunderland Ltd**

Vosper Thornycroft Ltd **Richard Klinger Ltd**

G. A. Platon Ltd **Taylor Pallister & Co Ltd**

PREFACE

This book on Engineering Drawing was compiled with two objects in view—to assist students studying for the Engineering Drawing Examination, set by the Department of Trade for a Second Class Certificate of Competency, and also to be of benefit to those studying for the Engineering Knowledge paper in Part B of the examination, as many of the examples used as drawing exercises appear in the syllabus in the latter examination.

In the drawing examination, the requirements of the DoT are somewhat different from most other examining bodies, and differ somewhat from standard drawing office practice. In order to ascertain the engineering knowledge of a candidate, a general assembly drawing is required, the details of which are given in the form of dimensioned pictorial views of the individual components, comprising some piece of marine engineering machinery. His skill as a draughtsman is judged from the attempt he makes at the drawing. It is expected that the particular piece of machinery could be manufactured from the drawing, and this necessitates inserting dimensions on a general assembly drawing, a practice not common elsewhere. As a result of this, many dimensions will have to be inserted on the actual components, ignoring the general rule of using projection lines to keep dimensions off components. To use projection lines for all dimensions on a general assembly drawing would make the drawing somewhat obscure.

Apart from this deviation the rules and conventions used in BS 308:1964, with amendments published in 1969, have been adhered to, and some of the more important of these have been quoted in the text.

The examinations set by the Department of Trade are in SI units, and as far as Engineering Drawing is concerned, the use of SI units, means that all dimensions will be stated in millimetres, a practice used throughout in this book.

With each pictorial representation, suggestions are given as to how to set about the orthographic projections required, with a suggested order with which to proceed. In addition, from the Engineering Knowledge point of view, in those examples where it is considered necessary, some explanation is given on the functioning of that particular piece of machinery. Candidates presenting themselves for the examination for a Second Class Certificate will have had some engineering experience, and a basic engineering training is assumed.

The question set in the Engineering Drawing examination is usually of such a nature as to suit either steam or motor candidates, from a point of view of assembling the components in order to attempt the general assembly drawing required. Each of the principal components on the question paper is labelled with a suitable material, and a list of these should be made in tabular form to complete the drawing. The scale to which the drawing is to be made is given in the question.

Drawing Instruments Required

The following drawing instruments will be required:

One pair of compasses with extension to draw circles up to about 300mm diameter.
One pair spring bow compasses.
One pair dividers.
2 set squares 45° and 60°. For projection purposes these should be fairly large.

Protractor.
Scale in metric units.
2 drawing pencils, H and 2H.
Rubber and eraser shield.
Fine ball point pen.
Drawing board and T square. These are supplied in the Examination, but for practice in drawing a student will require these.
Drawing clips (drawing pins should not be used).

Good draughtsmanship can only be achieved by practice, and students, especially those who have not done any drawing before, are advised to attempt some exercises at sea, even though they intend following a course of study ashore, in preparation for the examination. The examples on pages 20-32 are of a preliminary nature, and if these are attempted in the order presented in the book, a student should then be able to progress to some of the remainder of the examples which are of examination standard.

It is advisable at the outset to simulate examination conditions as far as possible, and in this respect some sheets of A2 drawing paper (594mm x 420mm) should be taken on the voyage, in order that the drawings may be attempted on the size of paper supplied in the examination. (See note on page 6 regarding recommended scales.) Six hours are allowed for the Engineering Drawing Examination, and students should bear this in mind when attempting those examples of examination standard. It is advisable to stop drawing 15 or 20 minutes before the end and insert dimensions, as without these a drawing would probably lose a considerable number of marks.

H.G.B.
Southampton, June 1977

CONTENTS

Chapter One	Hints on Producing a Drawing in Orthographic Projection	3
Chapter Two	Representation of Common Engineering Drawing Terms	8
Chapter Three	Examples in First Angle Orthographic Projection from Pictorial Representations	19
Chapter Four	Statics—Shearing Force and Bending Moment Diagrams	157
Table	Dimensions of Hexagon Nuts and Bolts	169
Index		170

Chapter One

HINTS ON PRODUCING A DRAWING IN ORTHOGRAPHIC PROJECTION

Engineering Drawing is a convenient means of conveying to others the ideas of a designer or draughtsman, in order that a component may be manufactured, or an assembly illustrated. Like all means of communication, certain rules and conventions are used, and these should be adhered to by the draughtsman, and understood by the person using the drawing.

Lines should be sharp and dense, and those specified as thick should be from two to three times the thickness of lines specified as thin. An important point to remember when drawing lines initially—do not lean heavily on the pencil—the weight of the pencil is sufficient, as it is much easier and quicker to go over the correct line than it is to rub out a line put in heavily.

Type of Line	Example	Application
Continuous (Thick)	———————	Visible outlines
Continuous (Thin)	———————	Dimension lines / Projection or extension lines / Hatching or sectioning / Leader lines for notes
Short Dashes (Thin)	— — — — —	Hidden details / Portions to be removed
Long Chain (Thin)	—— · —— · ——	Centre lines / Path lines for indicating movement / Pitch circles
Long Chain (Thick)	—— · —— · ——	Cutting or viewing planes
Continuous Wavy (Thin)	～～～～～	Irregular boundary lines / Short break lines
Ruled line and short zig-zags	—/\—/\—/\—	Long break lines

Types of Lines
As the line forms the most important part of an engineering drawing, the above types of line should be memorised and used where appropriate.

PROJECTION

Two systems of projection known as First Angle and Third Angle are in use in this country, and both are acceptable in the examination, so students who have studied engineering drawing before should use the system with which they are most familiar. The projection used should always be stated on the drawing. To avoid confusing one system with the other, however, First Angle will be used in this book. Should however the question specify the drawing to be made in Third Angle Projection the student should know the correct relative positions of the different views, and to illustrate this a comparison of both systems is shown on page 155. The Air Inlet Valve (page 94) shown in First Angle Projection on page 147 is drawn in Third Angle Projection on page 156.

First Angle Projection

In First Angle projection, each view shows what would be seen by looking on the far side of an adjacent view.

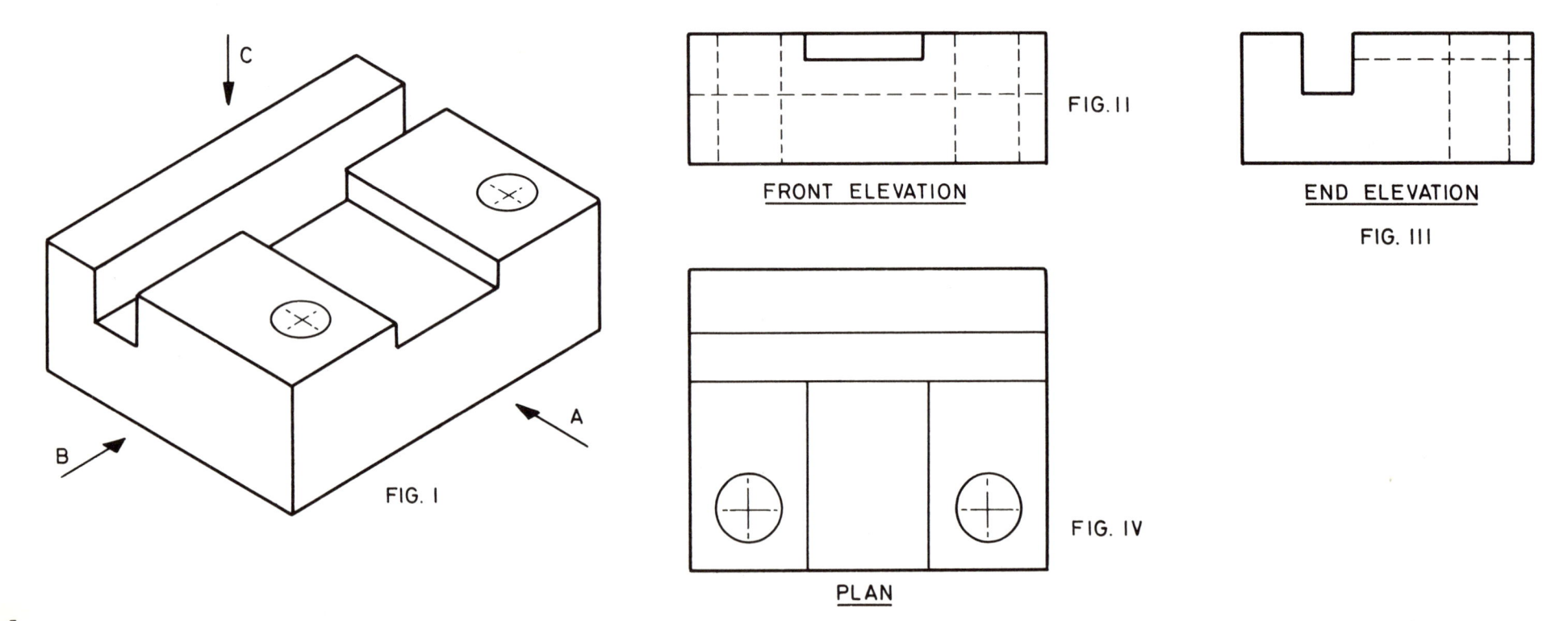

FIG. I

FRONT ELEVATION
FIG. II

END ELEVATION
FIG. III

PLAN
FIG. IV

A component is shown in isometric projection in figure I, and for this to be drawn in First Angle projection three views would be necessary. The front elevation in direction of arrow A would appear as in figure II, whilst the end view looking in direction of arrow B, would be as illustrated in figure III, drawn in line with the front elevation. The plan view, which is the view as seen when looking in direction of the arrow C, would appear as in figure IV drawn directly beneath the front elevation. The short dashes in the front elevation and end elevation, represent the hidden detail of the slots and holes.

SECTIONAL VIEWS

In general assembly drawings especially, there is normally too much hidden detail to represent by dotted lines, and for clarity sectional views are used. These are views representing that part of a component which remains, after a portion is assumed to have been cut away and removed. The exposed cut surface should then be indicated by section lining (hatching). Section lining should be made with thin parallel lines usually drawn at an angle of 45°, and suitably spaced in relation to the area to be covered. If the shape or position of the section would bring 45° section lining parallel or nearly parallel to one of the sides, another angle may be chosen. In all views showing sections of the same part, the section lining should be similar in direction and spacing. The section lining of adjacent parts should be in different directions or of a different pitch. Where the insertion of dimensions etc. on a sectional area is unavoidable, section lines should be interrupted.

HALF SECTIONS

Objects which are symmetrical about a centre line may be drawn having one half an outside view, and the other half in section. Hidden detail behind the cutting plane should generally be omitted, and should it be necessary to represent parts located in front of the cutting plane, these parts are represented by short thin chain lines. Exceptions to the general rule for indicating sections should be made where the cutting plane passes longitudinally through webs or ribs, shafts, bolts, nuts, rods, rivets, keys and pins. These should be drawn in outside view, and not in section.

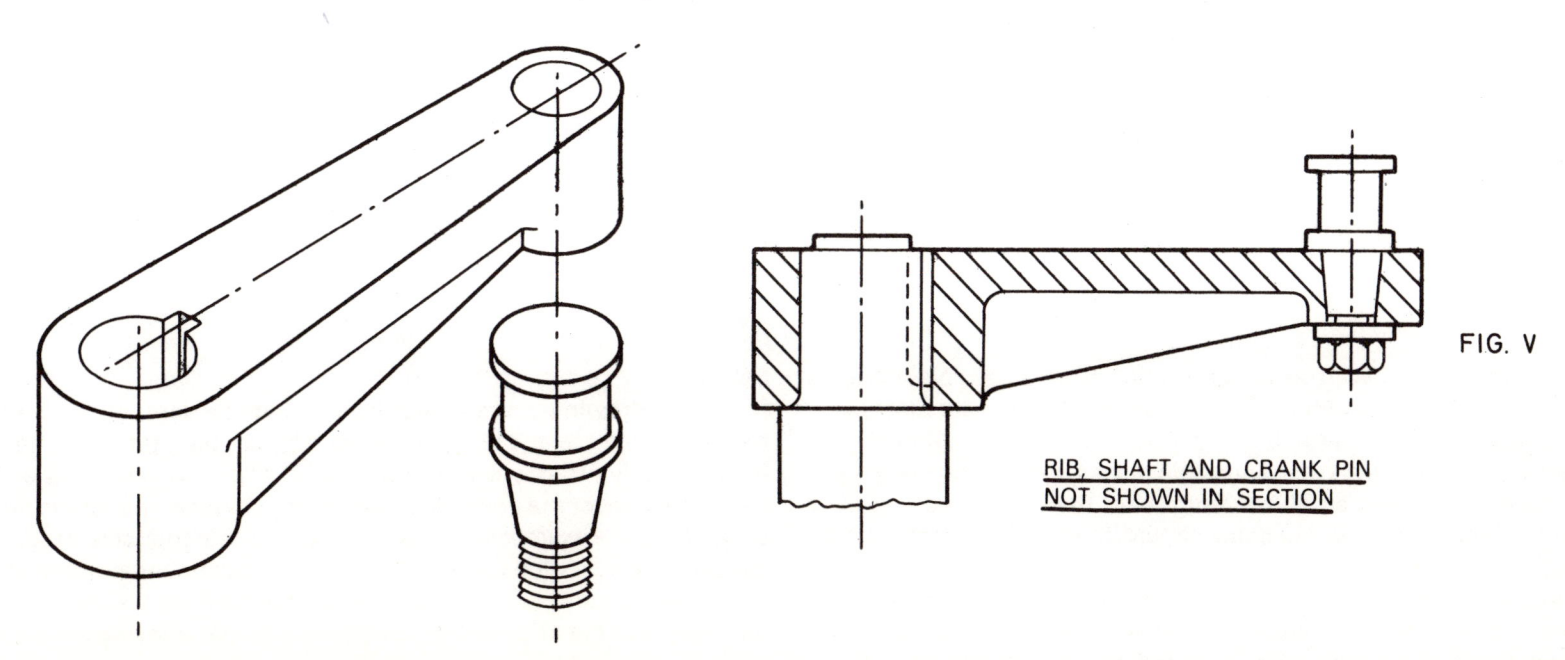

FIG. V

RIB, SHAFT AND CRANK PIN
NOT SHOWN IN SECTION

Fig. V shows a section through a crank, with the larger boss keyed to a shaft, and a crank pin fitted in a tapered hole in the smaller boss.

PROJECTION, DIMENSION LINES AND LEADERS

Projection lines are thin full lines, projected from points, lines, or surfaces to enable the dimensions to be placed outside the outline where possible. Where projection lines are extensions of lines of the outline, they should start just clear of the outline and extend a little beyond the dimension line. Dimension lines should be thin full lines, and where possible placed outside the outline of the object. Arrowheads should be about 3mm long, the point touching the projection or other limiting line. Dimensions should be inserted normal to the dimension line.

Figure VI shows the use of projection and dimension lines. Leader lines, used to indicate where dimensions or notes are intended to apply, should be thin full lines terminating in arrow heads. Instructions or notes should be underlined, and the leader line drawn from the end of the underlining to where the note applies. The latest edition of BS 308 recommends that instructions should not be underlined, but the drawings in this book were produced before this recommendation was published. Long leader lines should be avoided, even if it means repeating the dimension or note.

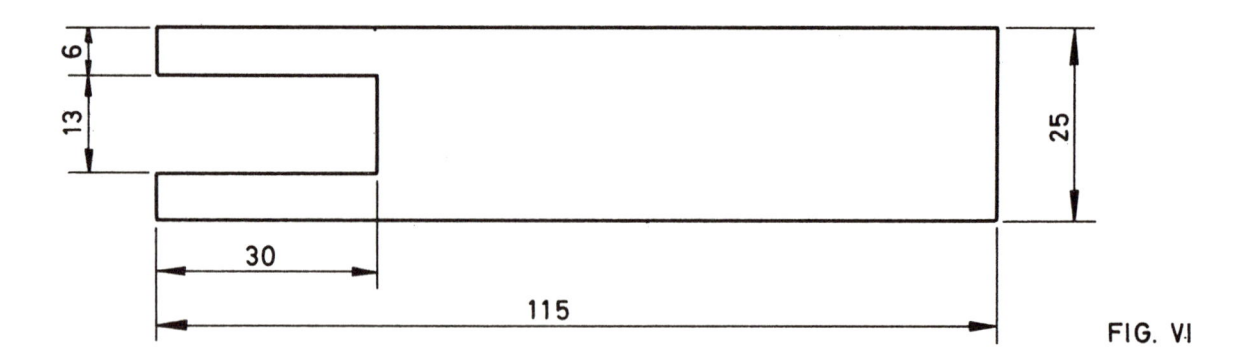

FIG. VI

SCALES

In metric drawings, scale multipliers and dividers of 2, 5 and 10 are recommended, and the scale used should be stated on the drawing. Representative fractions are written 1:1, 1:2, 1:5 and 1:10 etc., and this could come under the title at the bottom right hand corner of the drawing. The solutions to the examples in this book do not state the scale used, as with the comparatively small page size, the above recommendations with regard to scales would not permit utilising the page size to the full. This problem, however, will not arise when using standard A2 drawing sheets measuring 594mm x 420mm.

ABBREVIATIONS USED FOR GENERAL ENGINEERING TERMS

Term	Abbreviation	Term	Abbreviation
Across flats	A/F	Number	NO.
Assembly	ASSY	Pattern number	PATT NO.
Centres	CRS	Pitch circle diameter	PCD
Centre line	℄	Pneumatic	PNEU
Chamfered	CHAM	Radius (preceding a dimension)	R *
Cheese head	CH HD	Required	REQD
Countersunk	CSK	Right hand	RH
Countersunk head	CSK HD	Round head	RD HD
Counterbore	C'BORE	Screwed	SCR
Cylinder or cylindrical	CYL	Sheet	SH
Diameter (in a note)	DIA	Sketch	SK
Diameter (preceding a dimension)	Ø	Specification	SPEC
Drawing	DRG	Spherical diameter (preceding a dimension)	SPHERE Ø
Figure	FIG.	Spherical radius (preceding a dimension)	SPHERE R
Hexagon	HEX	Spotface	S'FACE
Hexagon head	HEX HD	Square (in a note)	SQ
Hydraulic	HYD	Square (preceding a dimension)	□
Insulated or insulation	INSUL	Standard	STD
Left hand	LH	Undercut	U'CUT
Long	LG	Volume	VOL
Material	MATL	Weight	WT.
Minimum	MIN	Taper, on diameter or width	▷

* CAPITAL LETTER ONLY

If the abbreviation forms another word i.e. fig. for figure, a full stop is placed after the abbreviation, otherwise a full stop is not required.

Chapter Two

REPRESENTATION OF COMMON ENGINEERING DRAWING TERMS

In this chapter, conventional methods of illustrating some of the more common items encountered in *Engineering Drawing* will be shown.

One of the most common items on an assembly drawing is a hexagon nut, and students at the beginning find difficulty in illustrating this. In the Whitworth series of hexagonal nuts, there was a simple formula for approximating the width across the flats, and width across the corners of the hexagon, in terms of the diameter of the bolt, but on studying the table of metric hexagons on page 167, it is obvious that no easily remembered formula can be devised. In the drawing examination tables are not supplied, so the student will have to approximate for hexagon sizes, if these are not given. A fair approximation could be taken as:—width across flats equals 1.5 times the diameter of the bolt, which would give a width across the corners equal to about 1.7 times the diameter. The thickness of the nut is approximately 0.8 times the diameter of the bolt, and the thickness of the bolt head about 0.7 times the diameter. Metric nuts have a 30° chamfer on both faces.

Example

Draw three views of an M20 bolt and nut, the bolt being 100 long and screwed for 30.

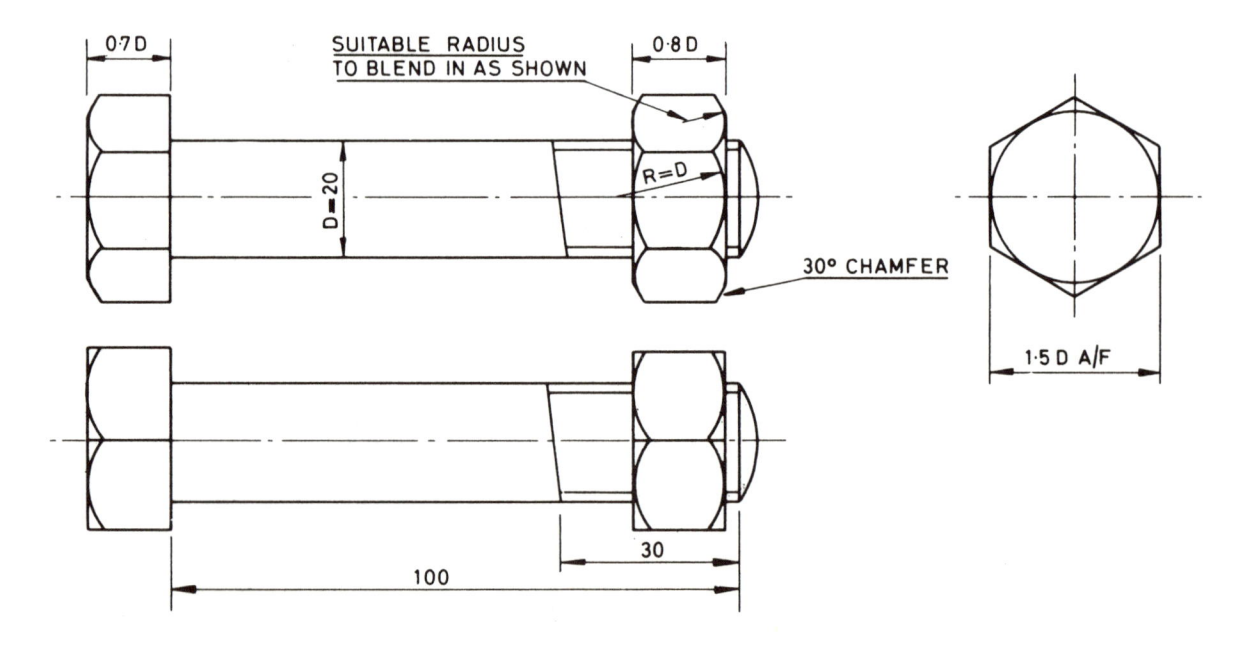

Conventional Representation of Common Features

TITLE	SUBJECT	CONVENTION	TITLE	SUBJECT	CONVENTION
EXTERNAL SCREW THREADS (DETAIL)			STRAIGHT KNURLING		
INTERNAL SCREW THREADS (DETAIL)			DIAMOND KNURLING		
SCREW THREADS (ASSEMBLY)			SQUARE ON SHAFT		
COMPRESSION SPRINGS			HOLES ON CIRCULAR PITCH		
TENSION SPRINGS			HOLES ON LINEAR PITCH		

TITLE	SUBJECT	CONVENTION	TITLE	CONVENTION
BEARINGS			WORMWHEEL (DETAIL)	
SPLINED SHAFTS				
SERRATED SHAFTS			WORM (DETAIL)	
BREAK LINES	ROUND (SOLID) / ROUND (TUBULAR) / RECTANGULAR		WORM AND WORMWHEEL (ASSEMBLY)	

10

TITLE	CONVENTION	TITLE	CONVENTION
SPUR GEAR (DETAIL)	ALTERNATIVE	BEVEL GEAR (DETAIL)	
SPUR GEARS (ASSEMBLY)		BEVEL GEARS (ASSEMBLY)	

11

The Ellipse

Occasionally a drawing appears in the Second Class Certificate examination where the construction of an ellipse is necessary. There are various methods of drawing an ellipse approximately, and one of these methods is given below.

major axis of the ellipse. The minor axis will be 30, the diameter of the bar. Draw two concentric circles of diameters equal to the major and minor axes respectively, and divide these circles into say 12 equal parts, by radial lines

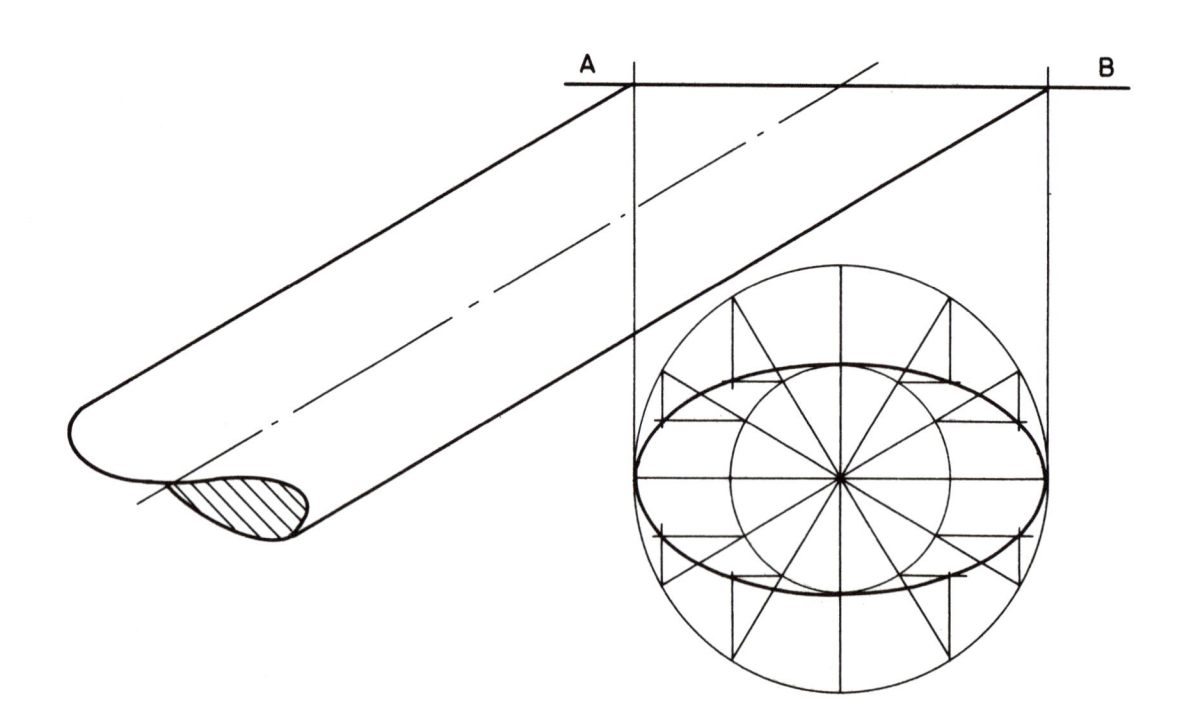

Example
The longitudinal axis of a cylindrical bar of 30 diameter is at 30° to the horizontal. Draw a section of the bar made by a horizontal plane.

The required section is an ellipse, and its construction is as follows. Draw the bar in elevation inclined at 30° to the horizontal, and let AB be the plane of the section. Project from A and B to the plan view underneath, and this will be the

drawn from the centre. This step is carried out with a 60° set square. Where these radial lines meet the circumferences of the two circles, draw vertical lines from the outer circle, and horizontal lines from the inner circle, and the point where these lines intersect will be a point on the ellipse. Draw a fair curve through these points.

The Involute

An involute could be defined as the locus of a point on the end of a piece of string, as it is held taut and unwound from a cylinder. Amongst its more common applications in engineering are the curved flanks of the teeth in spur gearing, and the profile of the casing of a centrifugal pump.

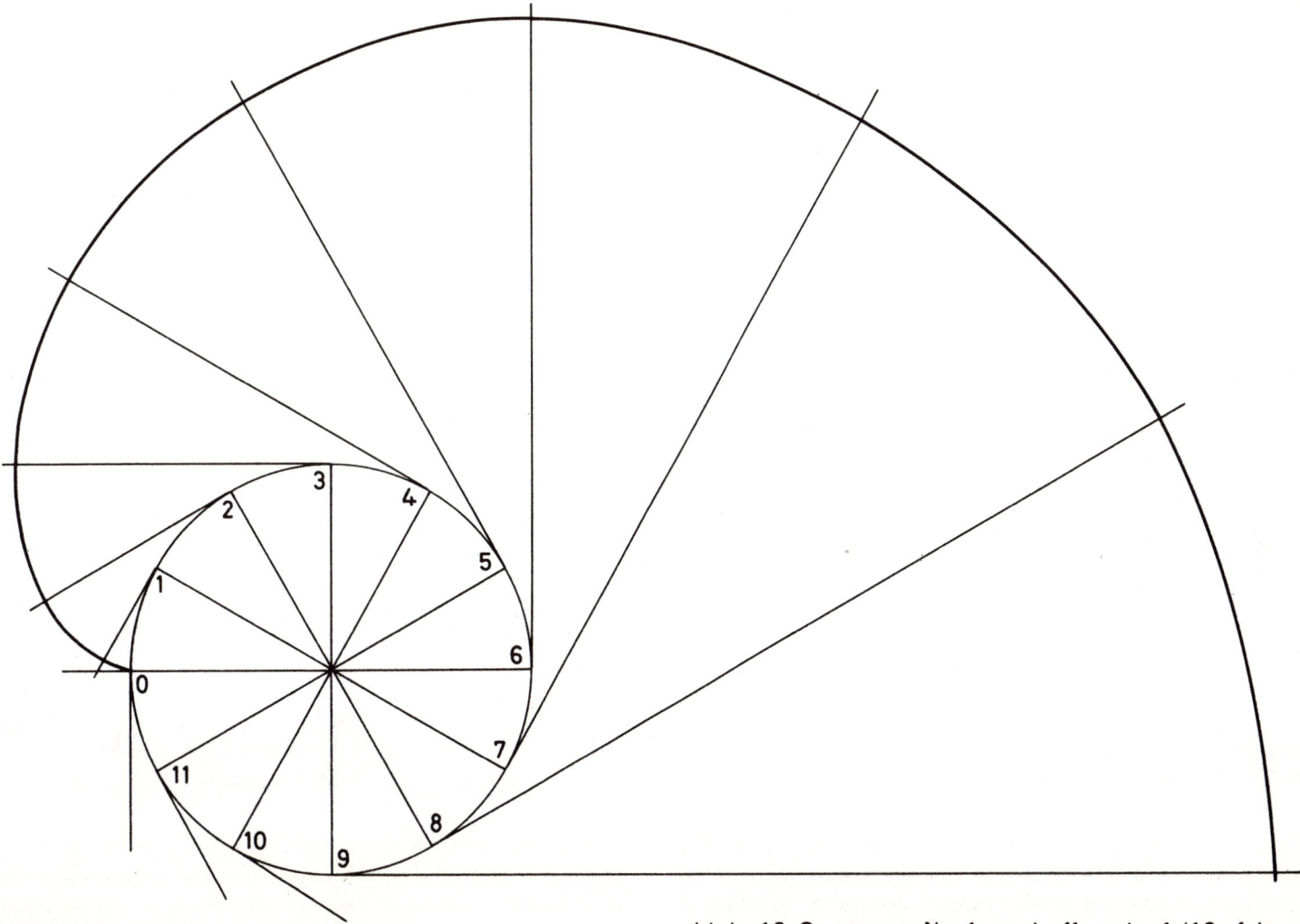

To Construct An Involute

Draw in the generating or base circle, and divide this into, say, 12 equal parts by using a 60° set square. Mark these radial lines 1, 2, 3 etc., and draw tangents to these. (Note that a tangent to a circle at a given point is at 90° to the radius at that point). Calculate the circumference of the circle, and divide this by 12. On tangent No. 1, mark off a point 1/12 of the circumference from the point of tangency, on tangent No. 2, mark off a similar point 2/12 of the circumference, and continuing in this manner a point will be obtained on each tangent. These points will be on the required involute.

Construction of a Helix

EXAMPLE:
CONSTRUCT A LEFT HANDED HELICAL SPRING 76 OUTSIDE DIA. 32 LEAD, COILS 16 SQUARE

EXAMPLE:
CONSTRUCT A RIGHT HANDED SQUARE THREAD 76 OUTSIDE DIA. 32 LEAD

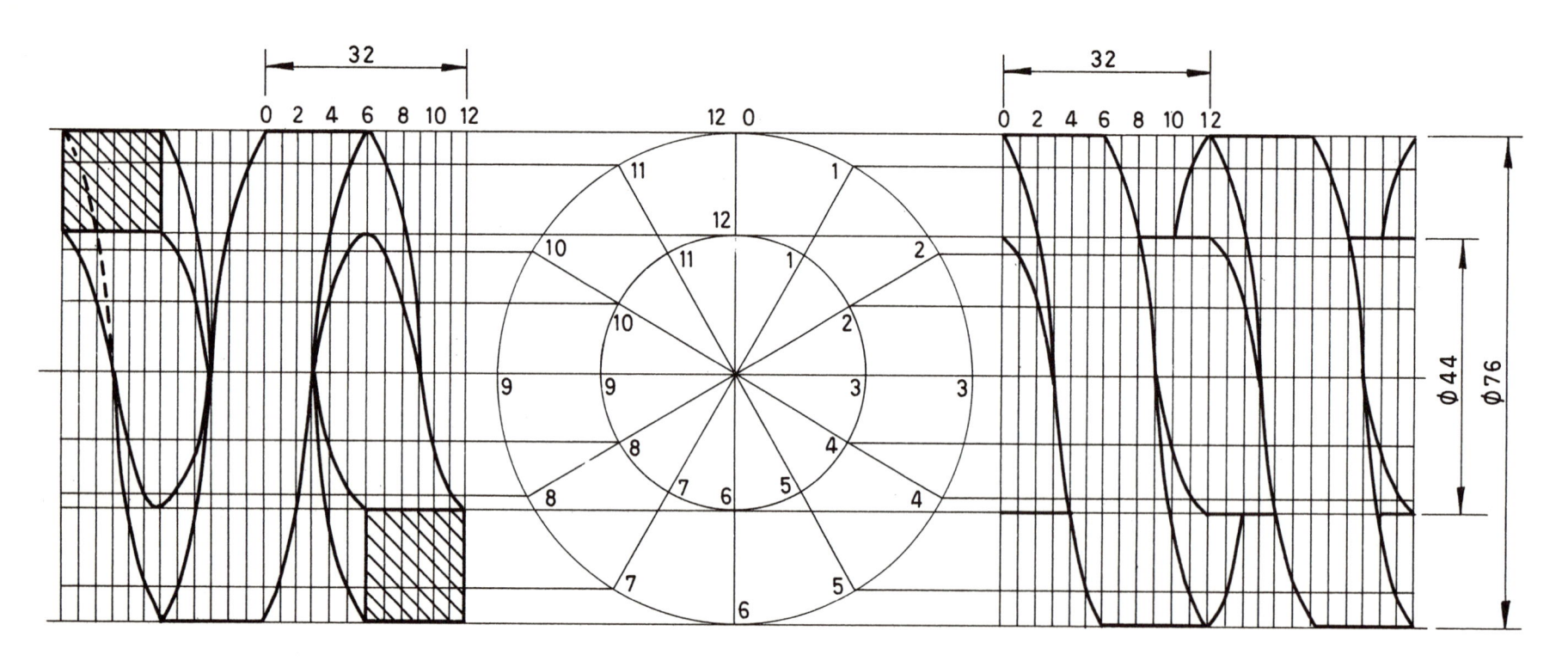

Construct concentric circles of 76 dia and 44 dia to represent the outside and root diameters respectively, and divide these into 12 equal parts, and number them. Project to the right for, say, the square thread, and to the left for the spring. Mark off the lead of 32 in each case, dividing this also into 12 equal parts, and draw vertical lines at each division. These should also be numbered as in the illustration. Project point 1 from the outer circle to vertical line 1 in the elevation for both screw and spring, point 2 to vertical line 2, and in a similar manner each point on the outer circle to its corresponding vertical line. These points will be on the required curve, and another curve drawn parallel to this at a distance of 16, will represent the other edge of the helix. In a similar manner, the points on the inner circle are projected to their respective vertical lines in both elevations, to give the helices of the root dia of the screw, and the internal diameter of the spring.

Curves of Interpenetration

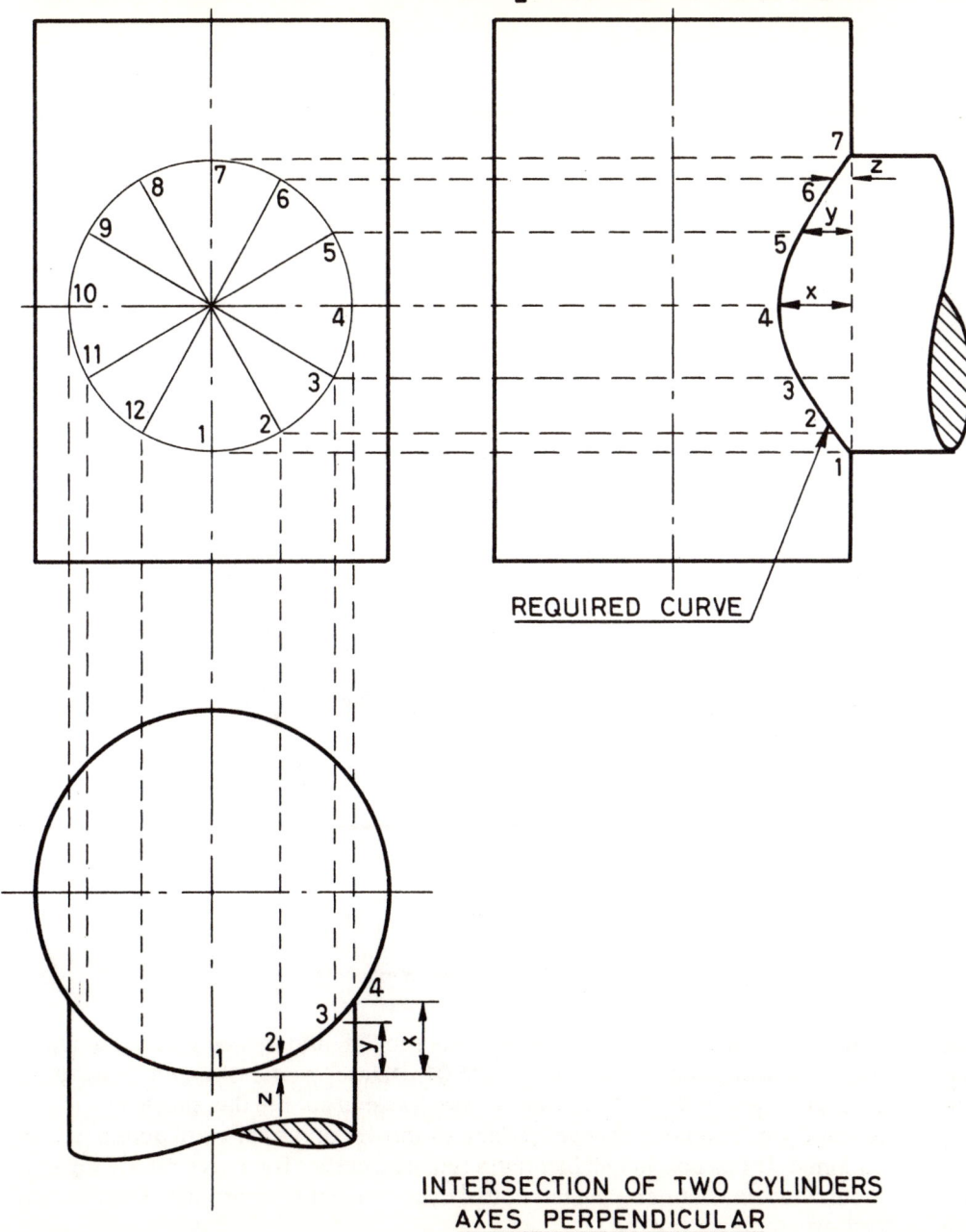

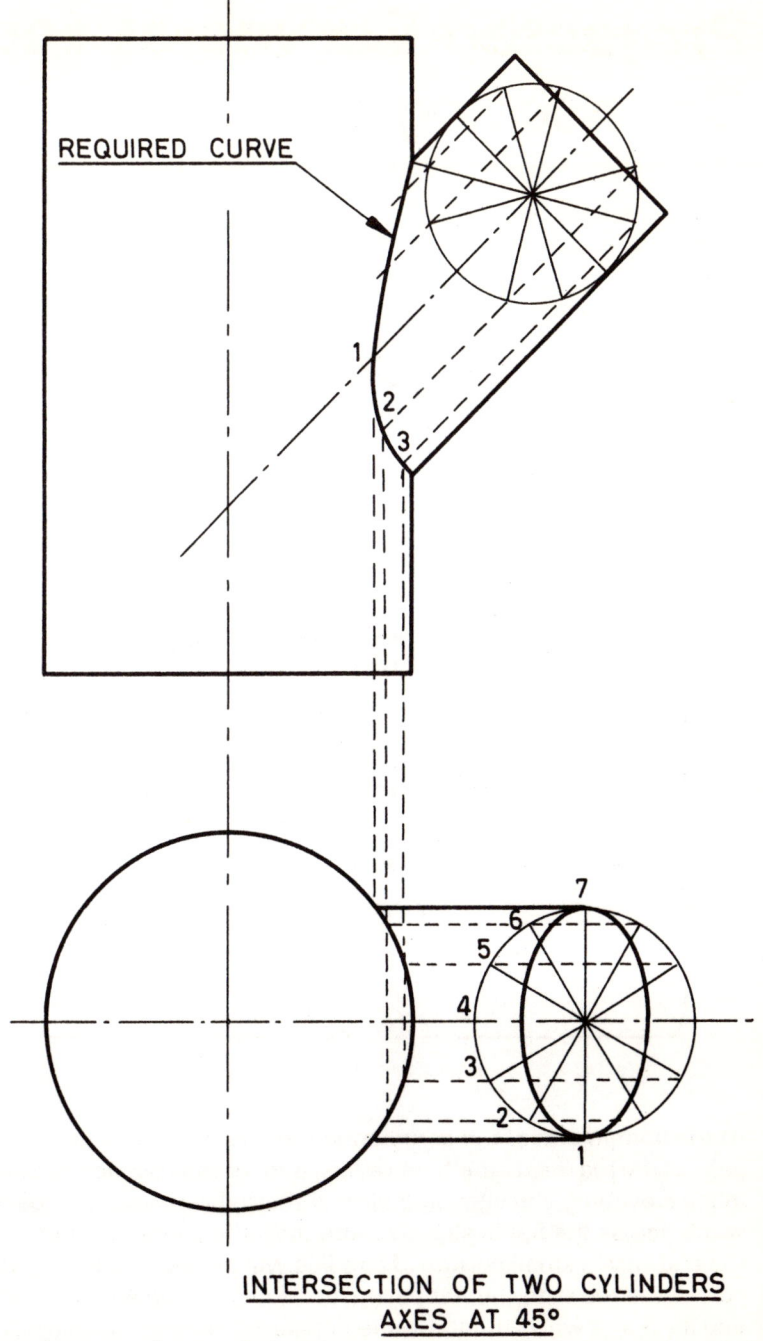

INTERSECTION OF TWO CYLINDERS
AXES PERPENDICULAR

INTERSECTION OF TWO CYLINDERS
AXES AT 45°

15

Projected Curves at top and bottom ends of Connecting Rod

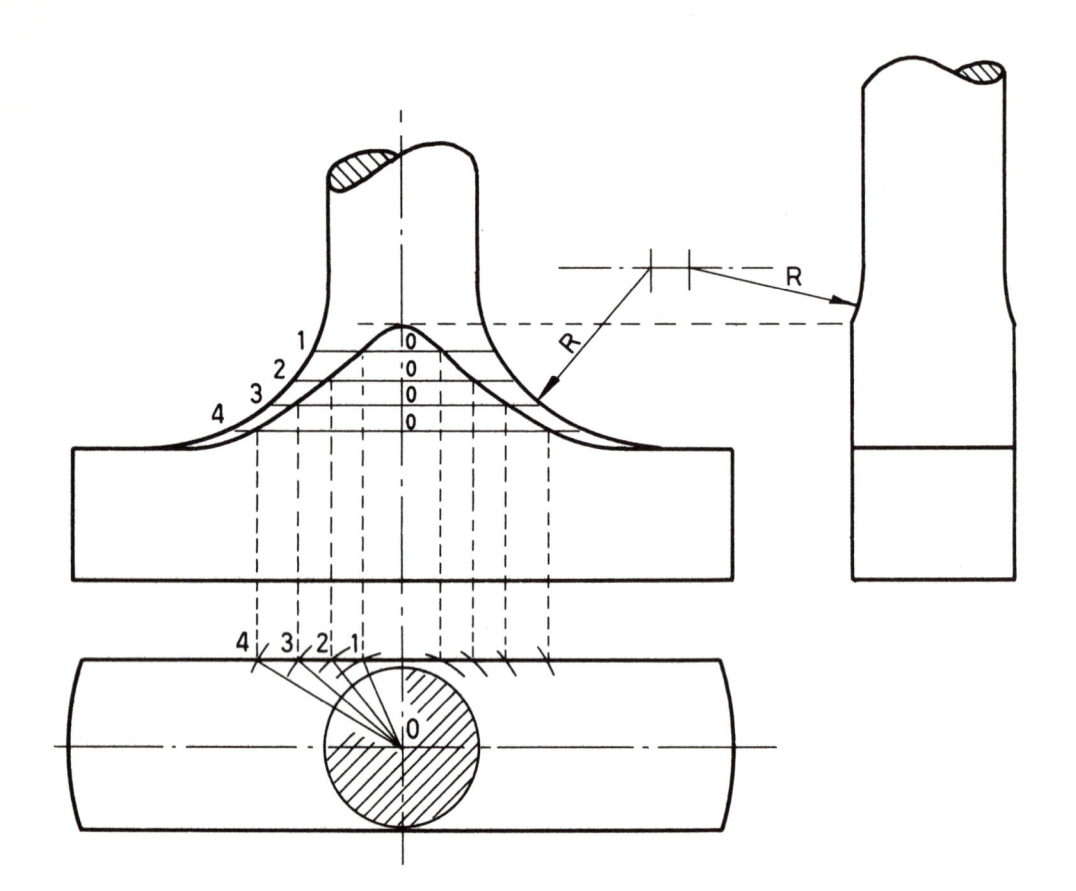

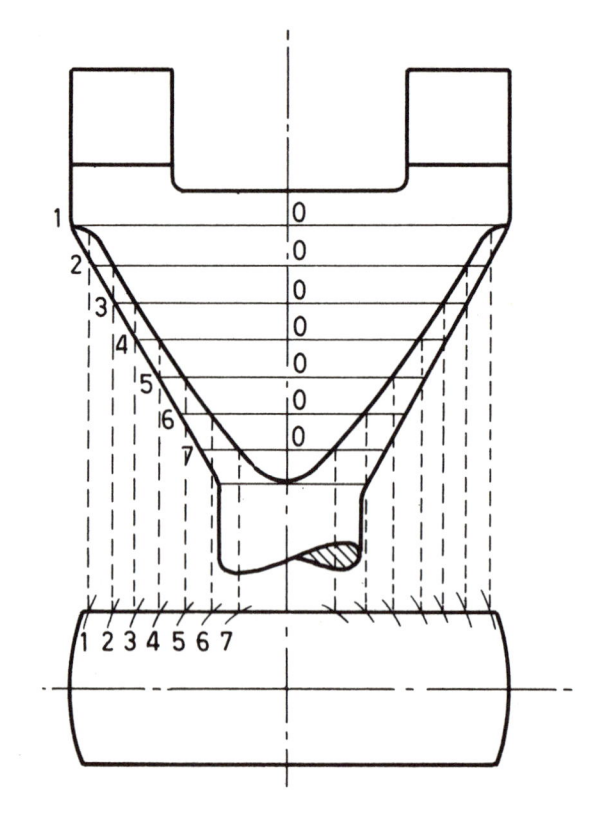

At the bottom end of the connecting rod a curve will be seen, where the flat palm of the rod meets the fillet radius, and its construction is as follows. Draw in the elevation, end view and plan of the rod as shown. In this example, the width across the flat is slightly more than the diameter of the rod, so that in the end view a small portion of a radius will be seen, marrying the cylindrical part of the rod with the flat part. This radius will be equal to the fillet radius, and its centre will be on the same horizontal centre line. Project across to the elevation the point where this radius cuts the flat, and this will mark the top of the curve. Construct a number of radii 01, 02 etc., and in the plan view draw arcs with these radii. At the points where the arcs cut the flat, project up to the elevation, each point to its own radius, as shown, and a series of points will be obtained. These points will lie on the required curve. The curve on the top end is obtained in a similar way, but in this case the width across the flats is equal to the diameter of the rod.

Spur Gearing

Occasionally a drawing appears where spur gearing has to be represented, and students should be familiar with the nomenclature, basic proportions, and an approximate method of drawing the teeth. The most common form for the tooth flank is the involute, and when it is made in this form the gears are known as involute gears. The angle of the pressure line (usually 14½° or 20°) determines the particular involute of the flank.

The illustrations below explain the nomenclature used in involute gearing. In addition to the definitions on the sketches, the term module is used in metric gearing, instead of the term diametral pitch used in the Imperial system. The module, which is the reciprocal of the diametral pitch, is defined as

$$\frac{\text{PITCH DIAMETER}}{\text{NUMBER OF TEETH}} \text{ or } \frac{D}{N}$$

and therefore is expressed in mm. If information is not given to the contrary the addendum for 20° pressure angle involute teeth can be taken as being equal to the module, whilst the dedendum, which includes the clearance, is about 1.16 x module.

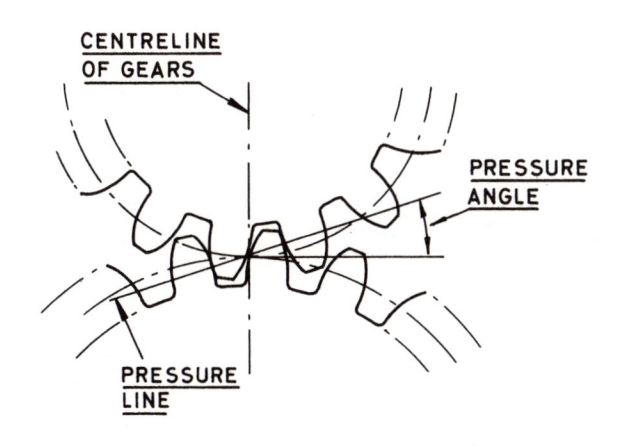

MATING GEAR TEETH. THE TOOTH PROFILES ARE INVOLUTE IN FORM AND ROLL ON EACH OTHER

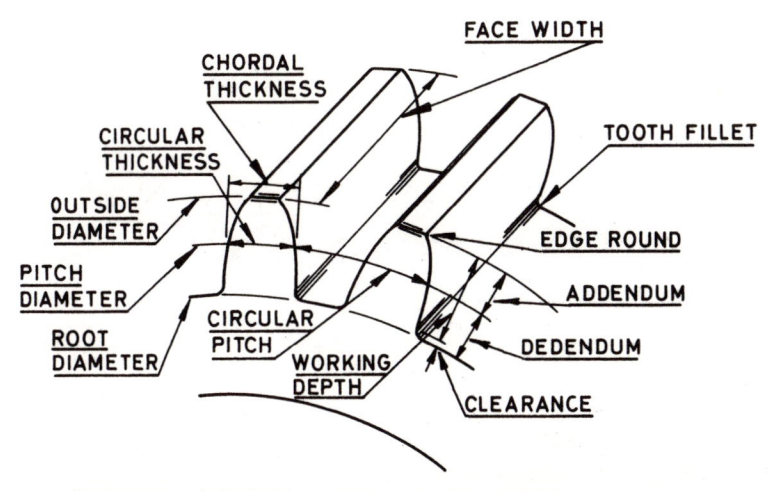

NOMENCLATURE USED IN GEARING

17

Approximate Method of Illustrating Involute Gearing

As the actual tooth profile illustrated on a drawing is of no importance in the cutting of the teeth, draughtsmen use an approximate circular arc method to illustrate the teeth. Working drawings of spur gearing do not generally include an illustration of the tooth profile, as the gear data given is sufficient for the machinist to cut the teeth, but occasionally students are required to show the approximate profile of a few teeth, usually to a larger scale than the main drawing.

There are a number of ways of approximating a tooth profile, and the method illustrated below gives a fair tooth form for a wheel with a large number of teeth. If the number of teeth is 20 or fewer, the radius used for scribing the arcs should be increased, or the appearance of excessive undercutting will be given. As will be seen from the diagram the teeth widths are marked off on the pitch circle, and the base circle on which the centres of the arcs are located is found by drawing a tangent to the pressure line, which passes through the pitch point. A small radius is used to marry the tooth flank with the root circle.

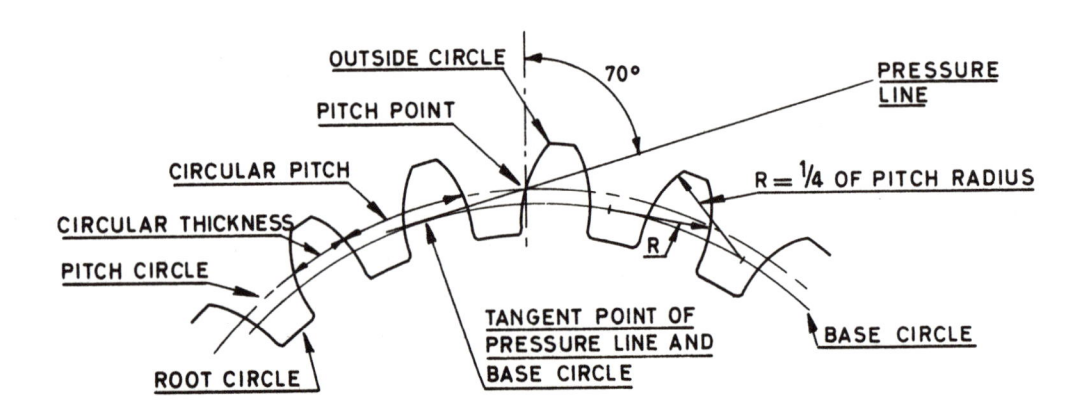

ACCEPTED METHOD OF ILLUSTRATING 20° PRESSURE ANGLE INVOLUTE TEETH. THE INVOLUTES ARE APPROXIMATED WITH ARCS OF CIRCLES, THE CENTRES OF WHICH ARE ON THE BASE CIRCLE

Chapter Three

EXAMPLES IN FIRST ANGLE ORTHOGRAPHIC PROJECTION FROM PICTORIAL REPRESENTATIONS

Title	Page
Air Inlet Valve	94, 147, 156
Automatic Valve	52, 126
Ballast Chest	42, 121
Bilge Suction Strainer	24, 112
Burner Carrier	56, 128, 58, 129
Centrifugal Brake	62, 131
Connecting Rod and Bearings	60, 130
Compressor Piston and Suction Valve	104, 152
Control Valve	30, 115
Crane Hook	28, 114
Crosshead and Guide Shoe	44, 122
Cylinder Relief Valve	26, 113
Feed Check Valve	46, 123
Flow Regulator	88, 144
Fuel Valve	90, 145
Fuel Control Lever	22, 111
Full Bore Safety Valve	86, 143
Gauge Glass (Plate Type)	106, 153
Gear Pump	48, 124
High Lift Safety Valve	84, 142
Hydraulic Steering Gear	102, 151
Machining Fixture	20, 110
Machined Block	20, 110

Title	Page
Main Gear Wheel	100, 150
Mechanical Lubricator	108, 154
Michell Thrust Block	98, 149
Oil Strainer	36, 118
Parallel Slide Stop Valve	38, 119
Pedestal Bearing	78, 139
Piston (4 Stroke)	70, 135
Piston (Upper and Rod)	72, 136
Piston Type Stop Valve	40, 120
Quick Closing Valve	64, 132
Reducing Valve	68, 134
Rudder Carrier Bearing	66, 133
Sealed Ball Joint	32, 116
Starting Air Valve	54, 127
Starting Air Pilot Valve	50, 125
Sterntube and Tailshaft	96, 148
Telemotor Receiver	74, 137
Tunnel Bearing	92, 146
Turbine Flexible Coupling	80, 140, 82, 141
Universal Coupling	34, 117
Valve Actuator	76, 138

Machining Fixture

DRAW

(i) Front elevation in direction of arrow.
(ii) End elevation.
(iii) Plan view.

Three views of the machining fixture shown on the opposite page are required, and on the normal size paper there should be no difficulty in fitting these in to a scale of 1:1. Commencing with the front elevation, as the length is 180 and the height 120, a rectangle is drawn to these dimensions, and a line 20 up from the bottom will represent the thickness of the base. Project these lines across to the end view on the right, and draw in the L shape form 100 wide, both legs being 20 thick. The plan view, directly below the front elevation, will be a rectangle 180 x 100, with a line 20 down from the top to represent the thickness of the vertical leg. Proceed now to draw in the slots in the vertical leg in the front elevation, and also those on the base, which will be seen in the plan view. The ends of these slots are semi-circular of radius 8, and these end radii should be drawn before the straight lines representing the sides, as a neater blending of curve and straight line will be achieved if done in this order. The M12 holes will appear in the plan view as circles, a 12 dia. broken circle representing the outside dia. of the thread, whilst a full circle is drawn on the same centre to represent the root dia. Hidden detail is illustrated in each view by short dash lines, and the drawing is finished by inserting dimensions, after the style of the finished drawing on page 110.

Also illustrated on the opposite page is a machined block, three similar views of which are required, and these should be attempted in a similar manner. In the front elevation in this case, commence by drawing a vertical centre line which will help locate the slot and the vee in the centre of the block, and then draw a rectangle 120 long by 80 high, after which the slot, projections on each side, and vee can be drawn in from the dimensions given. Project these across to the end view, which will be 120 long. The plan view, directly under the front elevation, will be a square of 120 side, with suitable lines projected from the elevation to represent the slot and the projections on the sides of the block. As before, the hidden detail in each view is represented by short dash lines, and the drawing should be sufficiently dimensioned so that the component could be produced.

The solution to this example also appears on page 110

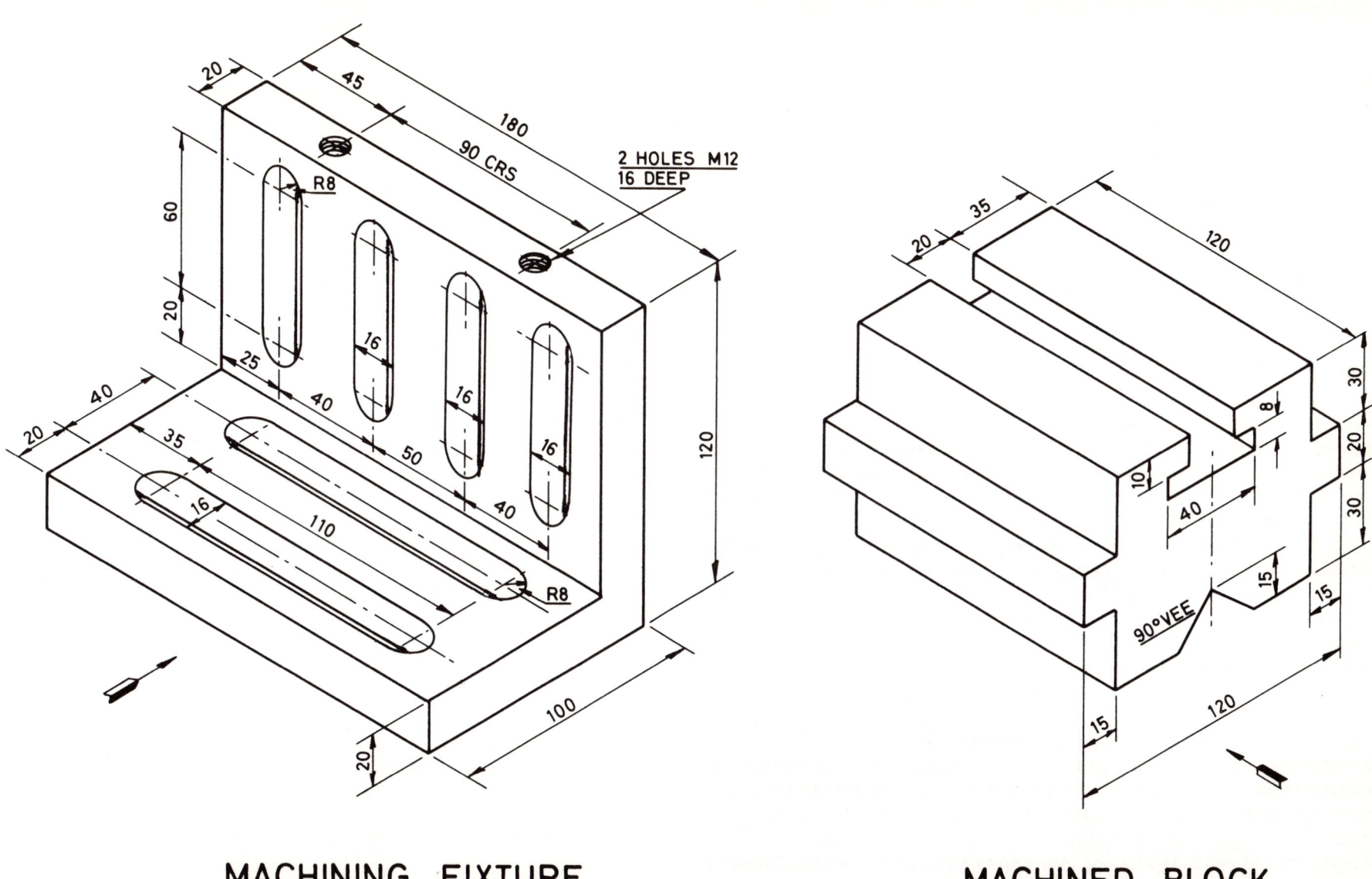

MACHINING FIXTURE

MACHINED BLOCK

21

Fuel Control Lever

DRAW

(i) Front elevation in section through axis of control shaft showing parts assembled.
(ii) End elevation.

A simple assembly drawing could now be attempted, and the Fuel Control Lever shown pictorially on the opposite page is included for that purpose. No difficulty will be experienced here in locating the parts, as it is obvious where each fits, and this drawing will show some of the components which are not shown in section.

Having decided upon a suitable scale, draw in a horizontal centre line for the fuel control shaft, and another 160 above this to indicate the centre line of the spherical stud, and produce these to the end elevation. Next draw a vertical centre line in the end elevation. In end elevation both bosses appear as circles, and these could be drawn in, and joined up tangentically. A circle 24 dia. in end view on the same centre as the large boss represents the end of the shaft and a web 10 thick equidistance each side of the centre line is then drawn in joining the two bosses. Project the boss circles over to the front elevation, and draw these in at their correct thicknesses. The 10 thick web is cast between the bosses, and this is now drawn in the sectional elevation. A 12 dia. clear hole for the stud is drawn on the centre line of the top boss, whilst a 24 hole is shown in the large boss enlarged to 36 dia. at the end of the boss to represent the fuel shaft. Draw in the key in end elevation, and project across to the front elevation to locate it in this view. The lever is 12 thick, and this is shown in section in the front elevation. Finally draw in the spherical stud and washer from the dimensions given, adding a standard M12 nut. Add section lining at 45° in the sectional elevation to indicate the cutting plane.

It should be noted that the following items on this drawing were not shown in section: shaft, key, spherical stud, washer and nut. As this is a preliminary drawing no material list has been added, but should the student wish to include this, the shaft would be of mild steel, the lever cast iron, the spherical stud of mild steel and the key of key steel.

The dimensioned drawing in orthographic projection appears on page 111

FUEL CONTROL LEVER

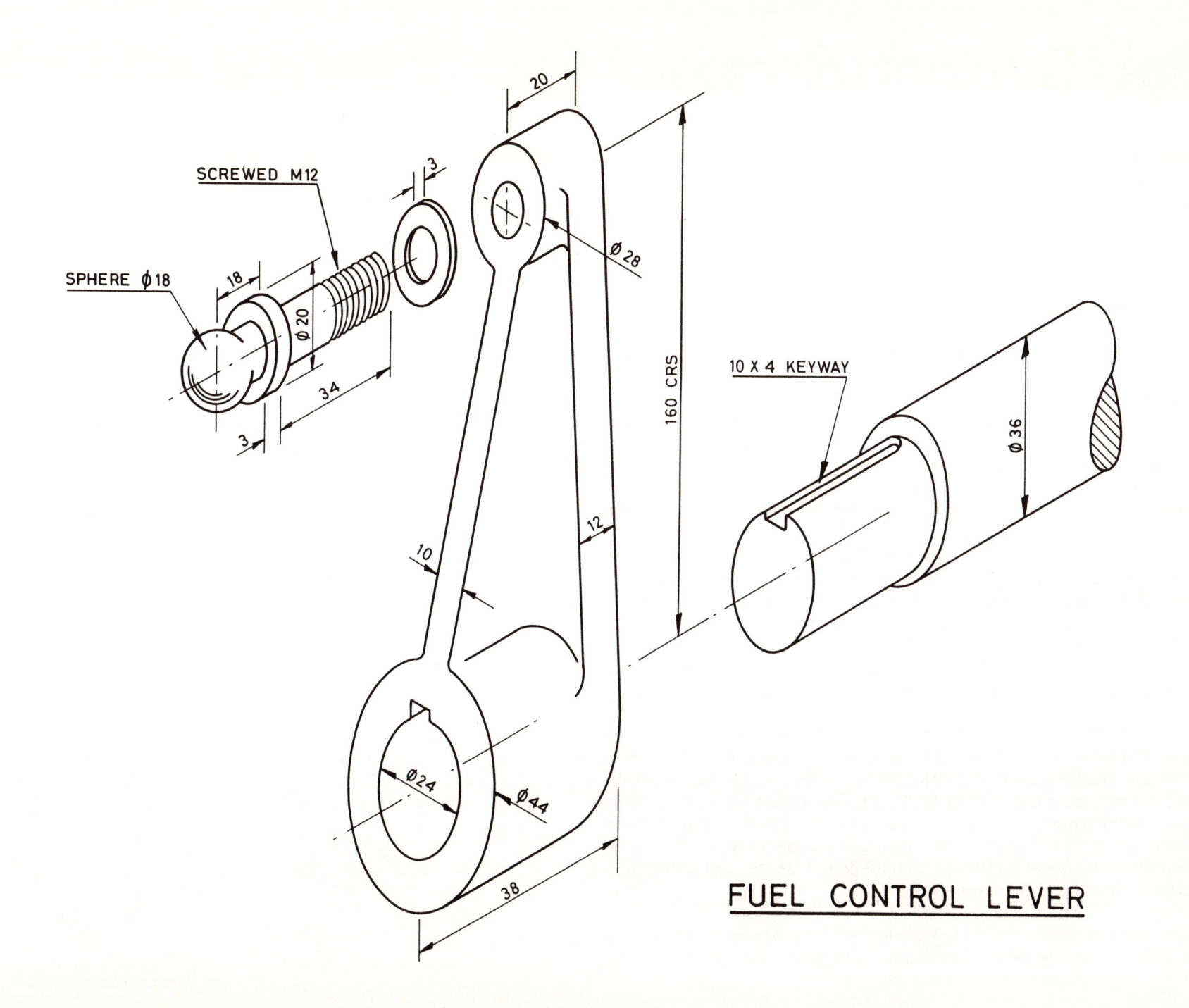

Bilge Suction Strainer

DRAW

(i) Sectional elevation through pillars showing all parts assembled.
(ii) End elevation.
(iii) Plan view.

INCLUDE ON DRAWING A LIST OF MATERIALS.

All marine engineers at an early stage in their sea-going career will have had an opportunity to examine the Bilge Suction Strainer illustrated pictorially on the opposite page. A high vacuum on the bilge pump suction but little or no water through the pump, usually indicates that the strainer in the strum box is choked, and it falls to him to clear the rags, waste, apple cores etc. etc. which have found their way into the bilges, and eventually into the strainer.

Having chosen a suitable scale, commence the sectional elevation by drawing in a vertical line to represent the face of the outlet flange. The left hand wall of the box is then drawn in 100 from this face. The overall length of the box is 266 + 2 x 14 = 294, so another line could be drawn in at this distance to represent the right hand wall. Before drawing in the horizontal line to represent the bottom of the box, work out the overall height from the bottom to top of M24 set screw as follows 14 + 175 + 68 + 37 = 294 to top of strong back. Allow say another 60 for the head of the set screw and a few threads for tightening purposes. The thickness of the walls is given as 14, so the inside of the casting could now be drawn in. The top flange, 25 deep and 25 from the walls is now drawn, and the lugs to take the M20 holes for the pillars. The centres of the pillars are found by subtracting 50 (2 x radius of the lugs) from overall length of strong back=410–50=360. The pillars shown full, cover (in section), strong back in section and M24 set screw shown full, could now be drawn in. The inlet branch, the centre line of which is 200 from the left hand wall of the box, and 55 down from bottom could now be drawn in. The outlet branch in this view could be drawn in chain dotted, as the section plane is behind the centre line of this branch.

The plan view will show the strong back and M20 set screw, cover with its stiffening webs, and also the outlet branch, the centre line of which is 33 below the horizontal centre line of the cover.

For completed drawing refer to page 112

BILGE SUCTION STRAINER

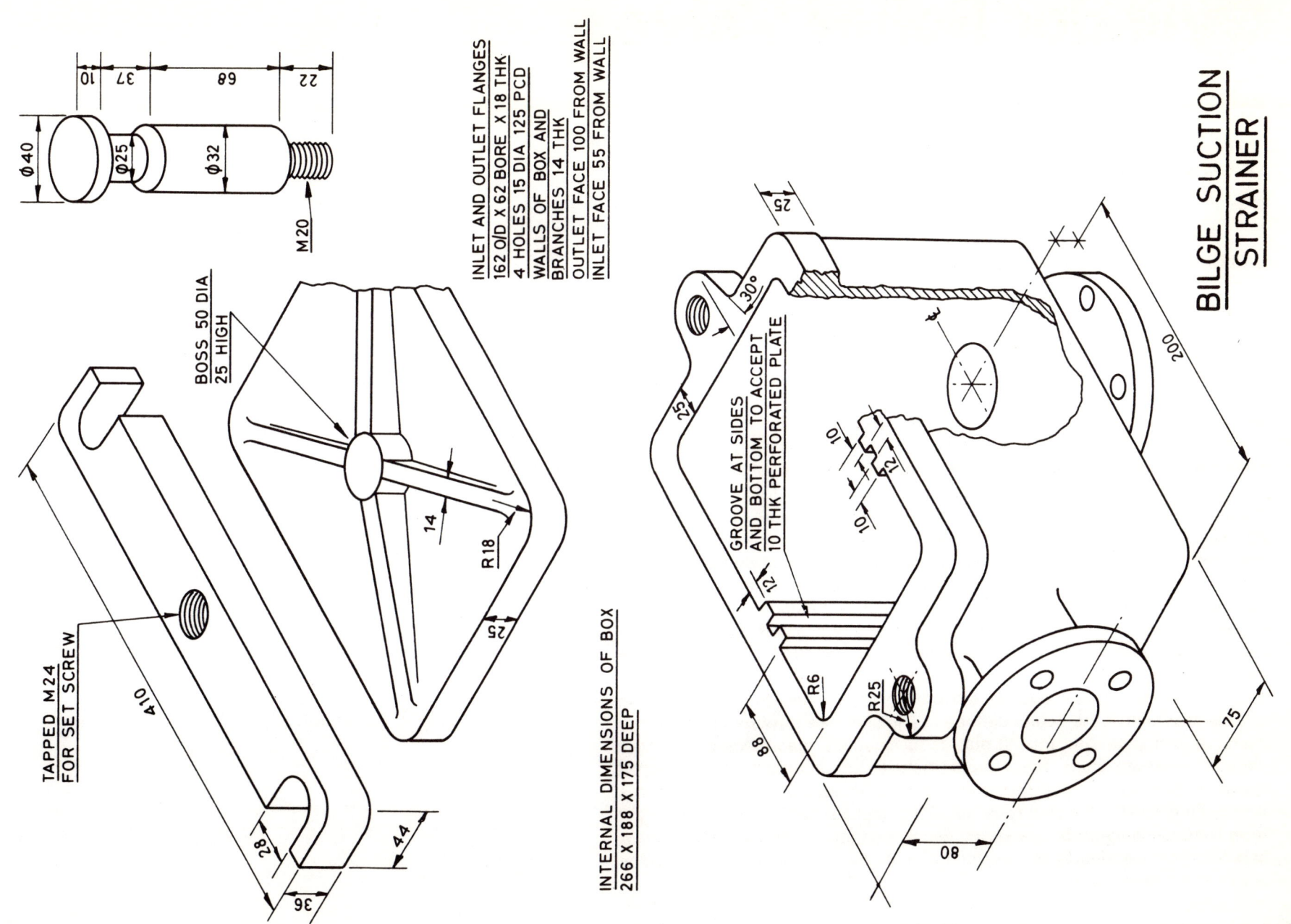

INLET AND OUTLET FLANGES
162 O/D × 62 BORE × 18 THK
4 HOLES 15 DIA 125 PCD
WALLS OF BOX AND
BRANCHES 14 THK
OUTLET FACE 100 FROM WALL
INLET FACE 55 FROM WALL

BOSS 50 DIA 25 HIGH

TAPPED M24 FOR SET SCREW

INTERNAL DIMENSIONS OF BOX
266 × 188 × 175 DEEP

GROOVE AT SIDES AND BOTTOM TO ACCEPT 10 THK PERFORATED PLATE

25

Cylinder Relief Valve

DRAW

(i) Elevation in section showing valve assembled.
(ii) End elevation (outside view).
(iii) Plan view (outside view).

INCLUDE A LIST OF MATERIALS ON THE DRAWING.

The pictorial drawing on the opposite page represents the components which comprise a Cylinder Relief Valve. Whilst this type of valve does not illustrate modern marine engineering practice, from a preliminary drawing point of view it is worth including.

Having decided upon a suitable scale for the size of paper, draw in two vertical centre lines for the sectional front elevation and plan, and also for the end elevation. These are used as datum lines for dimensions in a horizontal direction, and the base of the valve could now be drawn in, taking into account the overall height, and this line used as a datum line for vertical dimensions. At 162 up from the base line in both elevations, draw in the spherical radius of 42. On the sectional view where the thickness of the material will be seen, draw in another radius of 32 to represent the inside of the casing. The radii are now joined to the base by vertical lines, and radii at the bottom. Note that it is easier to blend a straight line with a radius, than a radius with a line. The base flange 114 square and 16 thick is now drawn in, and the 2 BSP tapped hole shown in the sectional view. Note that a 2 BSP thread measures about 57 dia. This is a relic of the past in dealing with pipe threads, where ¼in. was added to the tap size to give the actual diameter. The seat can now be added to the sectional elevation, and the 76 dia. flange which projects below the base shown in end elevation. The valve, which is not sectioned, is now added together with 2 coils of the spring in section, one at each end of its 146 free length. Finish off the top of the body to 38 dia. and a suitable radius, the height from the base being 225. The spring cap in section, and the adjusting screw and lock nut shown full, could now be drawn in, and the elevations finished by showing the escape ports.

The plan view consists of a 114 square to represent the bottom flange, with 4–14 dia. holes on a 114 PCD. No radius is given for the corners on the square, but as this is not important it is left to the discretion of the student. A circle 84 dia. represents the outside of the body, and a dotted circle 64 dia. the inside of the body, whilst two dotted lines 44 apart represent the ports. Finish off the plan view with a hexagon 30 A/F to represent the lock nut, concentric circles (the inner circle broken), and a square of 10 to illustrate the end of the 20 dia. adjusting screw.

For the completed drawing refer to page 113

CYLINDER RELIEF VALVE

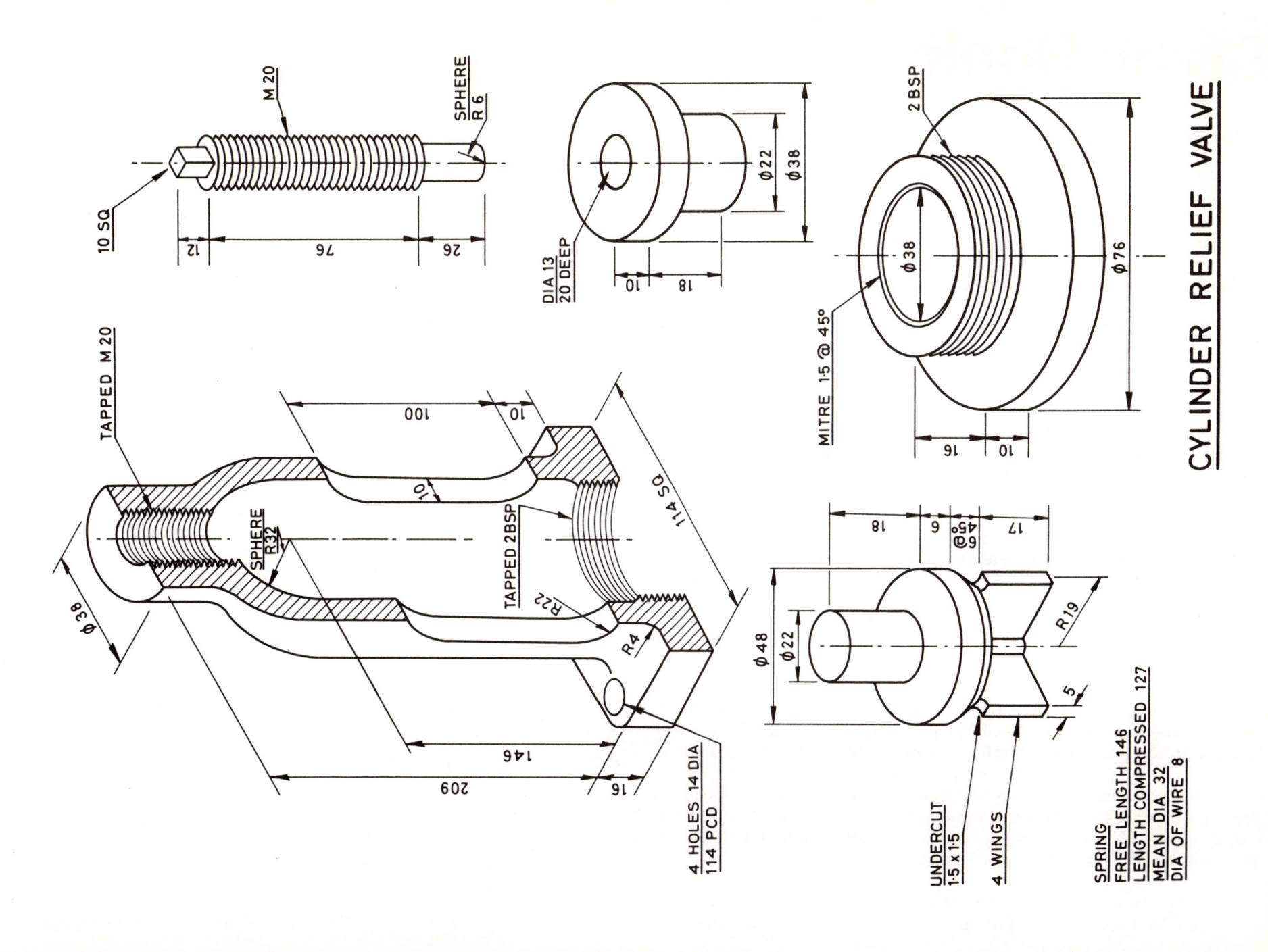

Crane Hook

DRAW

(i) Sectional elevation through the axis of the swivel block, showing all parts assembled.
(ii) End elevation.
(iii) Sectional plan view through axis of swivel block.

INCLUDE ON THE DRAWING A LIST OF MATERIALS.

Little assistance should be needed with the assembly of the Crane Hook shown pictorially on the opposite page, as it is a fairly common piece of equipment on board ship.

The side plates are held together by three 44 dia. studs, one of which is shown. The swivel block fits into two bushes, which are secured to the side plates by three M10 set screws in their flanges. The hook is free to revolve in a 40 dia. bush fitted in the swivel block, and a thrust washer is fitted between an M30 nut on the screwed end of the hook, and the top of the bush.

Leaving sufficient room for the sectional elevation, commence this exercise by drawing in the triangular-shaped side plate in end view, and mark off the centres for the three securing studs, together with the centre of the swivel block. The circular flange of the swivel block bush, three M10 set screws securing this bush, and four slotted nuts could also be drawn in.

For the depth of the side plates in the sectional elevation, project from the end view and draw these in 12 thick and 94 apart inside. Draw in the vertical centre line on this view, and again project across for the centres of the 44 dia. stud (only the top stud will be seen). The horizontal centre line of the swivel block is also obtained by projecting from the end elevation. Build up the swivel block around the centre line showing part of it in section, in order that the vertical bush may be shown. The hook could now be drawn in, noting that there is a clearance of 2 between the 60 dia. collar and the bottom of the swivel block. Finish off this view by adding the slotted nuts in section, the thrust washer and an M10 set screw in each bush flange.

The plan view taken through the horizontal centre line of the swivel block is now added, and the widths of the items in this view are obtained by projecting from the elevation immediately above it. The portion of the side plate seen in section in this view is found from the end view, by scaling off the distance from the vertical centre line, to where the horizontal centre line through the swivel block intersects the sides of the triangular side plate. The bottom two 44 dia. studs will also be seen in this view.

For the completed drawing refer to page 114

CRANE HOOK

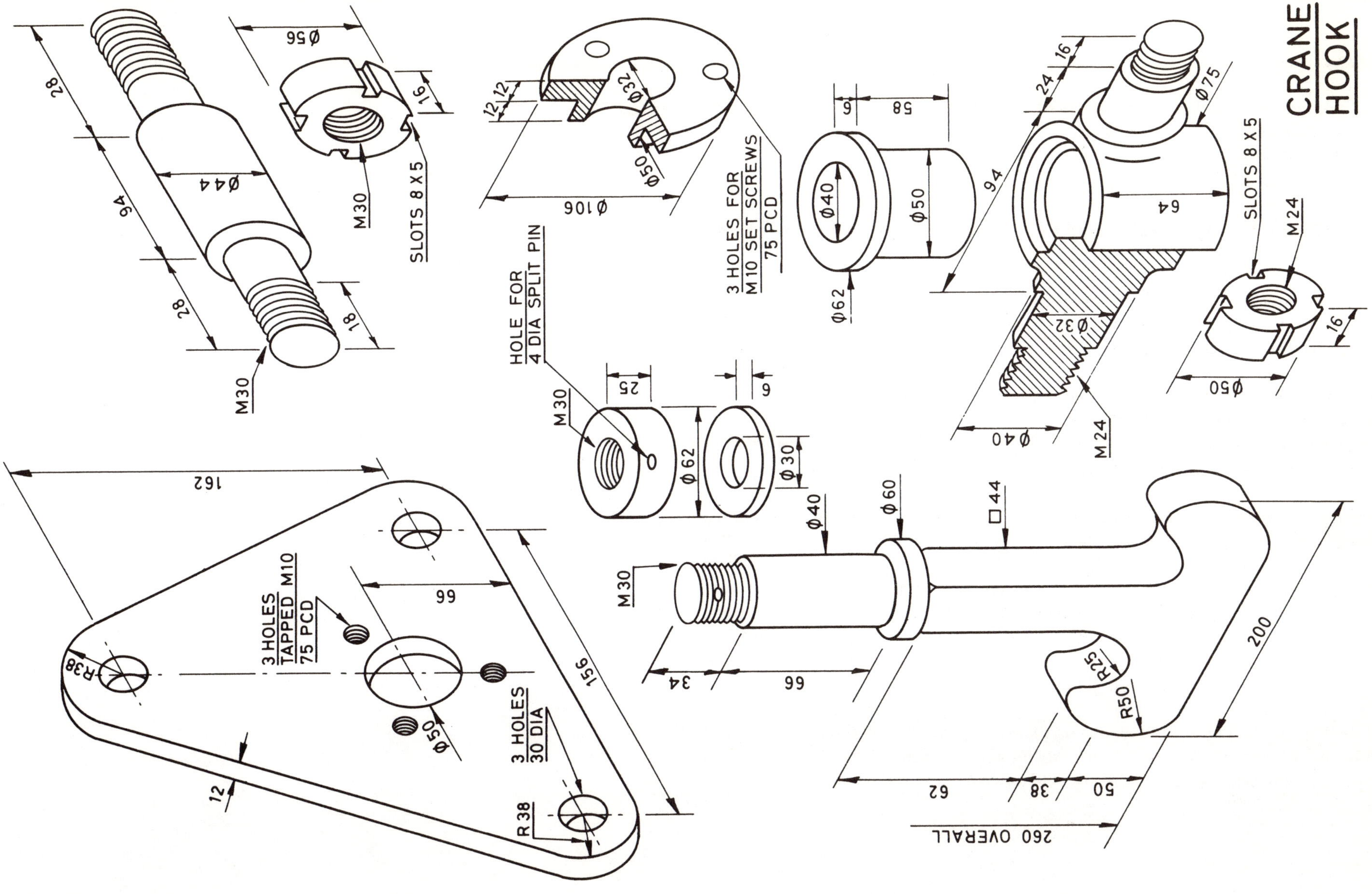
28
94
28
18
Ø44
M30
M30
Ø56
16
SLOTS 8 X 5
12
12
Ø32
Ø50
Ø106
3 HOLES FOR
M10 SET SCREWS
75 PCD
HOLE FOR
4 DIA SPLIT PIN
9
58
Ø40
Ø50
Ø62
16
24
Ø75
94
64
Ø32
Ø40
M24
SLOTS 8 X 5
M24
16
Ø50
25
6
M30
Ø62
Ø30
162
156
99
3 HOLES
TAPPED M10
75 PCD
R38
Ø50
3 HOLES
30 DIA
R 38
12
Ø40
Ø60
□44
M30
34
99
200
R25
R50
62
38
50
260 OVERALL

Control Valve

DRAW

(i) Longitudinal section through assembled valve with fulcrum pin vertically above right hand branch.
(ii) End elevation.
(iii) Plan view.

INCLUDE A LIST OF MATERIALS FOR A VALVE TO BE USED FOR LOW PRESSURE STEAM.

The control valve illustrated opposite could be used for a variety of duties where a simple automatic control is required. The operating lever would be connected at its free end to some type of actuator taking a pneumatic signal, and operating the valve accordingly. This valve could be used in the steam inlet to the heater of an air conditioning unit, the air pressure to the actuator being controlled by a thermostat in the room. It should be noted that as the valve opens downwards, loss of operating air pressure will allow the valve to close, giving it a fail safe shut characteristic.

For the sectional view required, draw in the horizontal and vertical centre lines, and produce to the end view and plan view respectively. From the dimensions given, draw in the valve body in section, omitting the horizontal division separating the inlet and outlet branches, until the valve seat is drawn in. As the valve seat locates on this division, it can now be drawn in at 6 thick below the seat. The valve spindle could now be drawn in, with the valve mitre locating on the seat mitre. Finish off the sectional elevation by drawing in the gland, fulcrum nut, and operating lever.

If the hexagons on the valve seat and gland nut are shown across the flats in the front elevation, they will, of course, be seen across the corners in the end view, and the widths across the corners can be found when the hexagons are drawn in plan view.

For the completed drawing refer to page 115

CONTROL VALVE

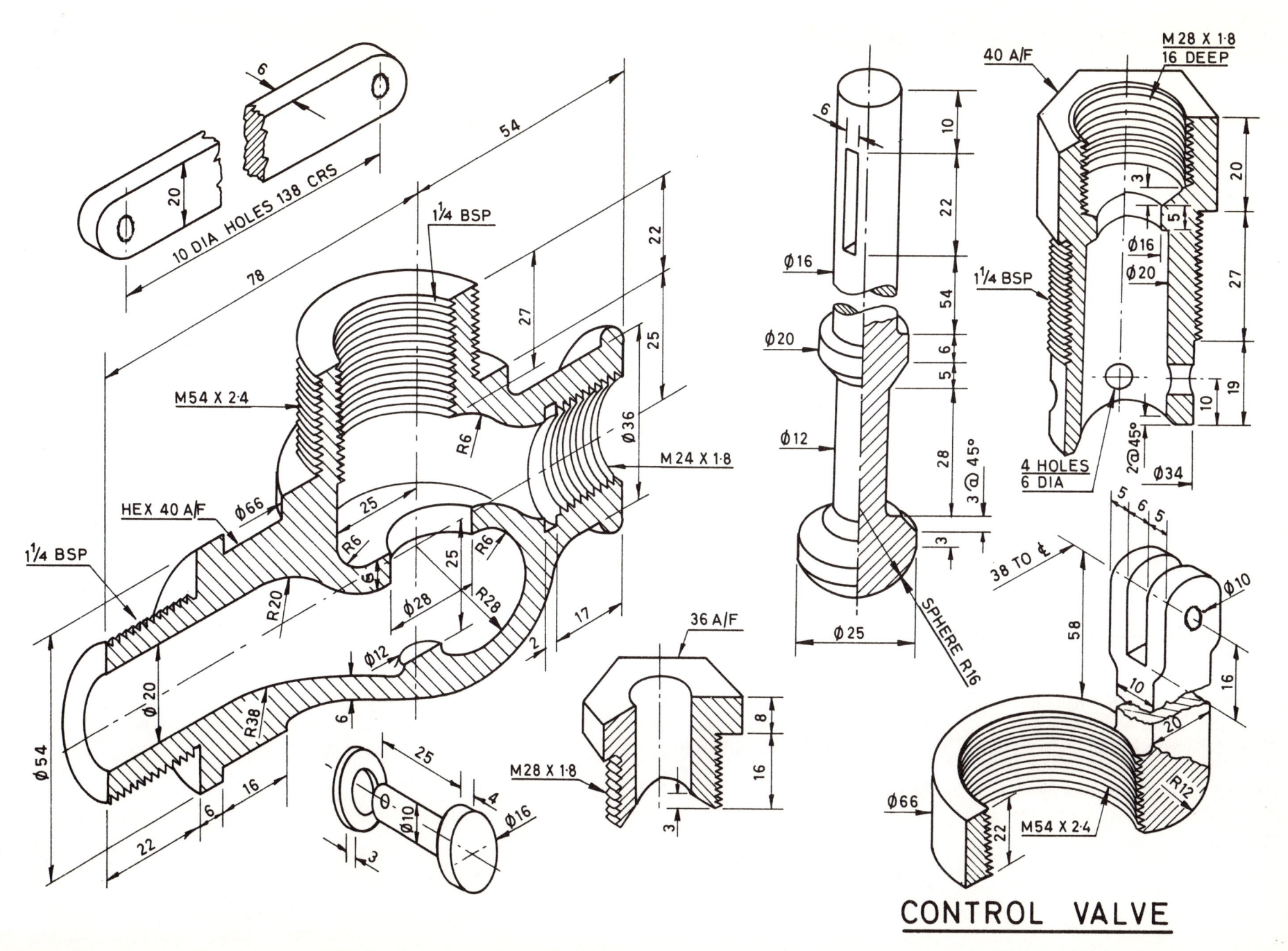

Sealed Ball Joint

DRAW

(i) Sectional elevation through the assembly.
(ii) End elevation.
(iii) Plan view.

INCLUDE ON THE DRAWING A LIST OF MATERIALS.

The sealed ball joint shown pictorially on the opposite page could be used in conjunction with a float, as a gas tight mechanism for a tank contents gauge. The body of the fitting is screwed 1¾ BSP to suit a tapped boss on the tank, and making a gas tight joint on the 56 dia. spigot, whilst that part of the lever in the tank is sealed from the atmosphere by a neoprene diaphragm, sandwiched between hemispheres on both the male and female parts of the lever.

Commence the sectional elevation by building the body of the fitting around its horizontal centre line. When the 2 thick diaphragm has been located between both parts of the body, the centre of oscillation of the mechanism can be marked, as this coincides with the centre of the diaphragm. On this centre, draw in the hemispheres on both levers, noting that 1 is machined off the flat faces of both hemispheres to accommodate the diaphragm. Finish off the levers from the details given, and add an M10 nut and lock nut to the inner lever, for the purpose of tightening the flat faces of the hemispheres against the diaphragm.

Six stiffeners 20 wide are provided on the left hand part of the body, to take the tapped holes for six countersunk screws holding both parts together, and these stiffeners will be shown radially in the end view. Their location from the ₵ in the plan view can be obtained by picking up the various edges by dividers from the vertical centre line in the end elevation.

For the completed drawing refer to page 116

SEALED BALL JOINT

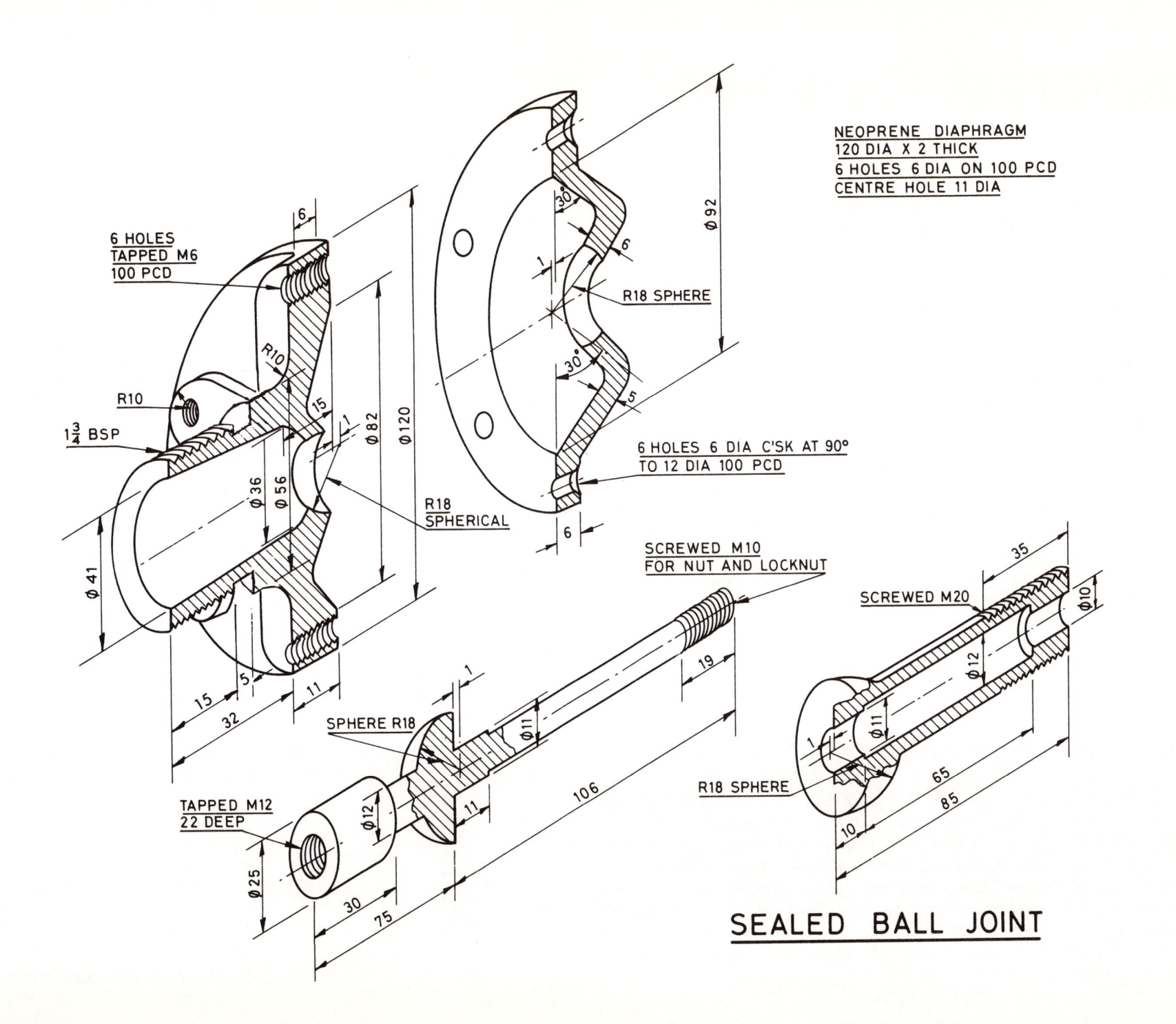

NEOPRENE DIAPHRAGM
120 DIA X 2 THICK
6 HOLES 6 DIA ON 100 PCD
CENTRE HOLE 11 DIA

6 HOLES TAPPED M6 100 PCD

R10
R10
1¾ BSP
R18 SPHERICAL
Ø36
Ø56
Ø41
Ø82
Ø120
6
15
15
5
11
32
1

Ø92
R18 SPHERE
30°
6
1
30°
5
6
6 HOLES 6 DIA C'SK AT 90°
TO 12 DIA 100 PCD

SCREWED M10
FOR NUT AND LOCKNUT
SPHERE R18
TAPPED M12
22 DEEP
Ø25
Ø12
Ø11
30
75
11
106
19
1

SCREWED M20
R18 SPHERE
Ø10
Ø12
Ø11
35
65
85
10
1

33

Universal Coupling

DRAW

(i) Elevation of assembled coupling, one half in section, other half an outside view.
(ii) End view, one half in section through swivel pins, other half an outside view.

INCLUDE ON THE DRAWING A LIST OF MATERIALS

The universal coupling illustrated opposite is of a rather more sophisticated design than the common type found on valve extended spindles, and could be used on a motor launch or a lifeboat propeller shaft, where the shaft was raked relative to the centre line of the engine in order to keep the propeller submerged.

The pictorial view does not indicate the number of each component required for the complete coupling, and the student from his engineering knowledge would have to know that two forked ends are required, the forks being similar to those on the centre forked link. Two coupling bodies would also be required to take the eight bushes for the eight pins: A Stauffer grease lubricator is fitted from end of one forked end to the centre of coupling

Before deciding the scale, work out the overall length as follows. Distance from end of one forked end to the centre of coupling

$$= 198 + \frac{225}{2} = 310.5$$

therefore the total length over the forked ends will be double this.

Draw the horizontal centre line, and then draw in the centre forked link making, say, the top half a sectional view. On the same vertical centre lines through the holes for the pins, locate the pin holes of the two forked ends, and draw these in from the details on the pictorial view. One pin, with its nut shown full, and bush shown in section, in each fork end should be drawn in on the sectional part of the view, and around these two bushes locate the two split coupling bodies 114 wide. On the bottom half of the elevation, show the 16 dia. holes, eight bolts securing each half of the housing. A lubricator on each pin would complete this view.

The end view with, say, the left hand side a section through the pins will show the coupling body or housing in more detail, with the horizontal and vertical pitches of the 16 dia. bolt holes. In this view one pin on the link would be seen full and two half pins in the forked ends.

For the solution to this exercise refer to page 117

UNIVERSAL COUPLING

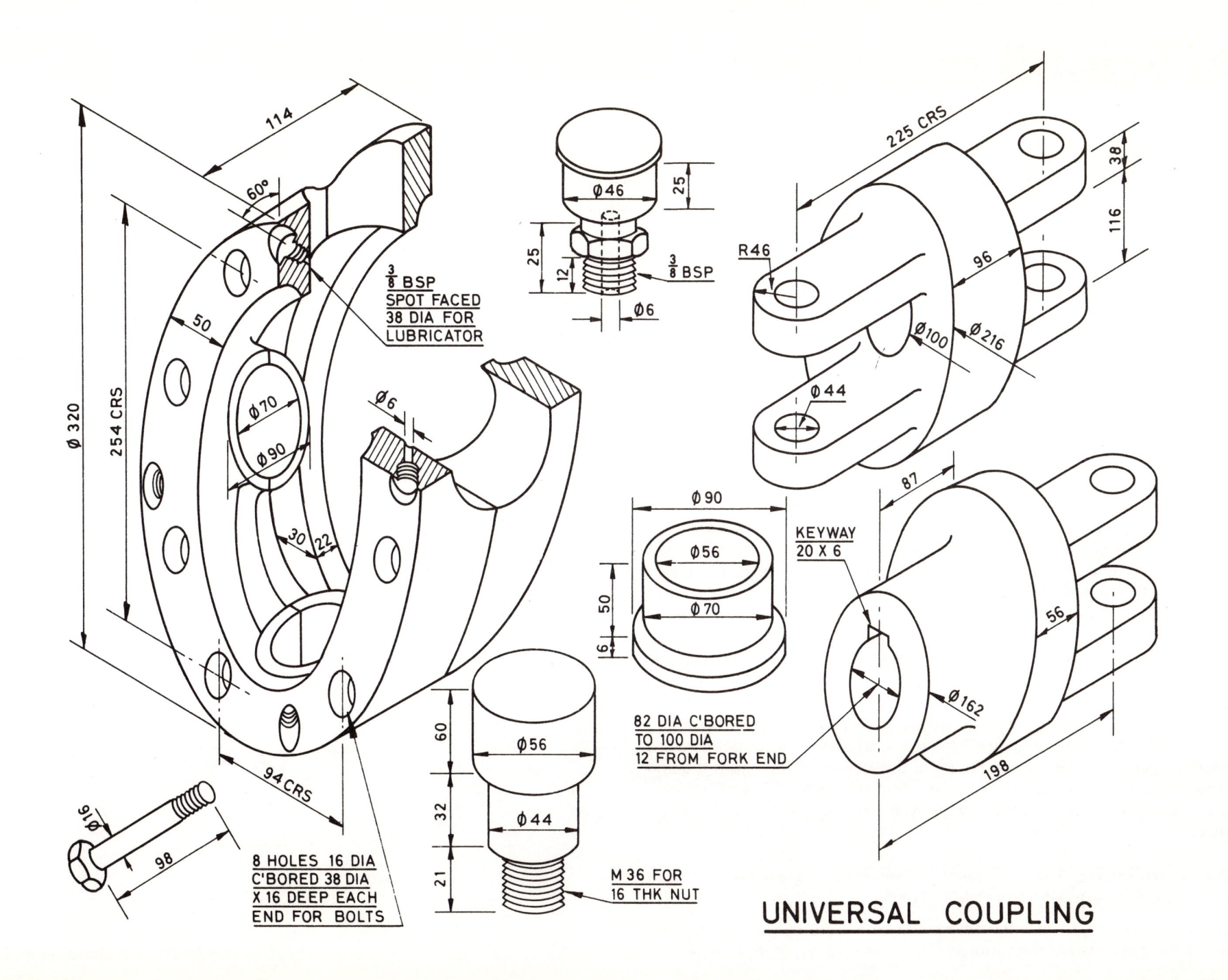

Oil Fuel Strainer

DRAW

(i) Sectional elevation through inlet and outlet branches showing all parts assembled.
(ii) End elevation.
(iii) Plan view.

INCLUDE ON THE DRAWING A LIST OF MATERIALS.

Marine Engineers should need little introduction to the Oil Fuel Strainer shown pictorially on the opposite page. On a steam ship, the strainer is usually fitted before the oil fuel unit, to protect the pump from large foreign matter which may have found its way into the settling tank, whilst another filter, generally of the knife edge type, is fitted after the heater. The strainer would be in duplicate, in order that one unit could be cleaned without shutting down the pump.

In the sectional elevation, draw vertical centre lines for both filter elements at 180 centres, and horizontal centre lines for the outlet and inlet branches at 204 centres, and produce the horizontal centre lines to the end view. Sufficient dimensions are given for the body of the strainer to be drawn around these. No dimension is given for the overall height of the strainer, but the bottom edge of the mounting feet would be slightly below the inlet flange. The division, separating the inlet side from the outlet side, has two bosses 162 dia. x 132 bore x 30 deep, into which fit the lower ends of the strainer elements. The upper flanges on these elements fit into the recessed holes in the top of the strainer, together with the spigots machined on the bottom of the top cover. Before drawing in the strongbacks in the sectional elevation, locate the pillars in the end elevation. These pillars to take the strongbacks are screwed into lugs cast on the top flange of the strainer body. From the end elevation project to the sectional view to obtain the height of the strongbacks, and show an M20 set screw in each, bearing on a 50 dia. x 5 deep boss cast on the cover.

The plan view is projected from the sectional elevation above it, and widths in this view picked up by dividers from the end view.

In the sectional elevation two curves are shown, where the inlet and outlet branches meet the body of the strainer. The construction for these curves of penetration is given on page 15, but if time does not allow for this construction, the extreme point on the curve in each case can be found by projecting up from the plan view and a smooth curve drawn in.

For the completed drawing refer to page 118

OIL FUEL STRAINER

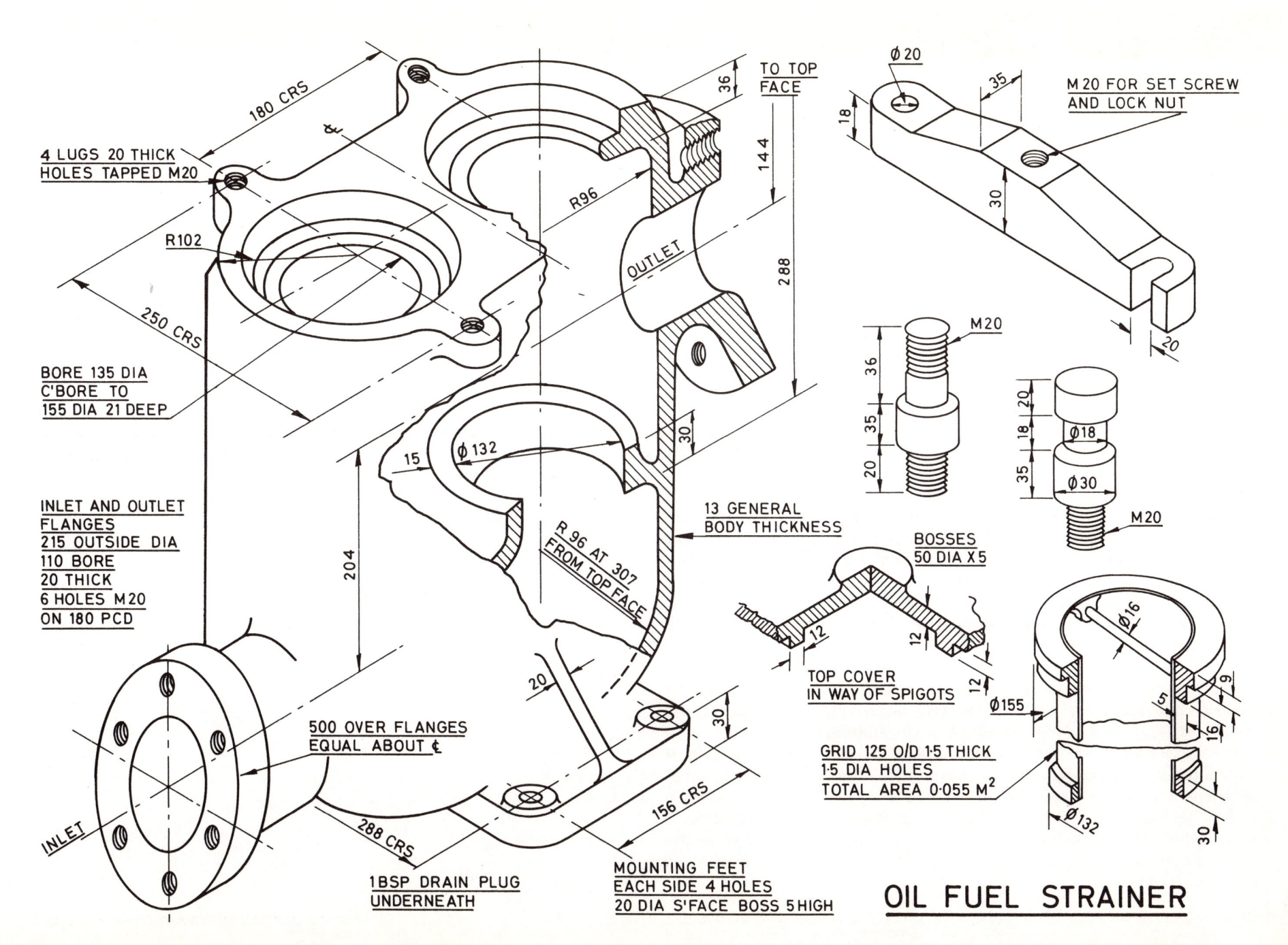

- 180 CRS
- 4 LUGS 20 THICK HOLES TAPPED M20
- R102
- R96
- 250 CRS
- 36 TO TOP FACE
- 144
- 288
- OUTLET
- BORE 135 DIA C'BORE TO 155 DIA 21 DEEP
- INLET AND OUTLET FLANGES
  215 OUTSIDE DIA
  110 BORE
  20 THICK
  6 HOLES M20
  ON 180 PCD
- 15
- Ø132
- 30
- 204
- 13 GENERAL BODY THICKNESS
- R 96 AT 307 FROM TOP FACE
- 20
- 30
- INLET
- 500 OVER FLANGES EQUAL ABOUT ₵
- 288 CRS
- 1 BSP DRAIN PLUG UNDERNEATH
- 156 CRS
- MOUNTING FEET EACH SIDE 4 HOLES 20 DIA S'FACE BOSS 5 HIGH
- Ø20
- 35
- 18
- M20 FOR SET SCREW AND LOCK NUT
- 30
- 20
- M20
- 36
- 35
- 20
- 20
- 18
- 35
- Ø18
- Ø30
- M20
- BOSSES 50 DIA X 5
- 12
- 12
- 12
- TOP COVER IN WAY OF SPIGOTS
- Ø16
- Ø155
- 9
- 5
- 16
- 30
- Ø132
- GRID 125 O/D 1·5 THICK
  1·5 DIA HOLES
  TOTAL AREA 0·055 m²

37

Parallel Slide Stop Valve

DRAW

(i) Elevation in section through branches, showing valve assembled.
(ii) End Elevation.

INCLUDE ON THE DRAWING A LIST OF MATERIALS.

Although the valve illustrated on the opposite page is a relatively small example of this type of valve, it is fairly common on board ship in bores up to 300 dia., and can be used on high pressure steam. When placed in a pipe line, the valve provides unrestricted passage through the valve equal to the bore of the pipe. Pressure drop is minimised, and the valve will pass a quantity of steam, or fluid, equal to the full carrying capacity of the pipe. Its chief drawback is maintenance in situ, but this is overcome by renewing the valve lids and seats. The seats have lugs cast on their bore (not shown on drawing), to accommodate a seat withdrawal tool.

Build-up the valve body in both views around vertical and horizontal centre lines, and draw in the valve seats in the sectional elevation. The valve spindle has a 50 dia. boss forged on its lower end, through which passes the female valve disc, and into this valve disc fits the male disc, both being pressed to their seats by a spring. It should be noted that the tension in the spring has nothing to do with keeping the valve tight when shut—it is fitted merely to keep the valve from collapsing.

The cap for the chest can now be drawn in both views, and the square threaded nut located by its 32 dia. collar at the bottom of the M36 hole for the stuffing box. This hole is 19 deep, so that with the stuffing box in place there is a clearance of 2 between bottom of stuffing box and collar on nut. Finish off the sectional elevation by drawing in gland and gland nut and finally the handwheel, which fits on the 12 square and secured by an M10 nut and washer. In the sectional elevation, the square threaded spindle nut could be broken in a similar fashion to the illustration in the pictorial view, in order to show the internal detail, and at the same time leave the square and M10 screwed portion as a full view.

If the hexagons on the cap, stuffing box and gland nut are viewed across the flats in sectional elevation, the widths across the corners in the end view can be found by:—

$$\text{Width across corners} = \frac{\text{Width A/F}}{\cos 30°}$$

This method is suggested as plan views of these items are not required, and, therefore, the widths cannot be projected, unless, of course, hexagons are drawn off the drawing, and the required widths measured. For suggestions on drawing a hexagonal nut see page 8.

For the completed drawing refer to page 119

PARALLEL SLIDE STOP VALVE

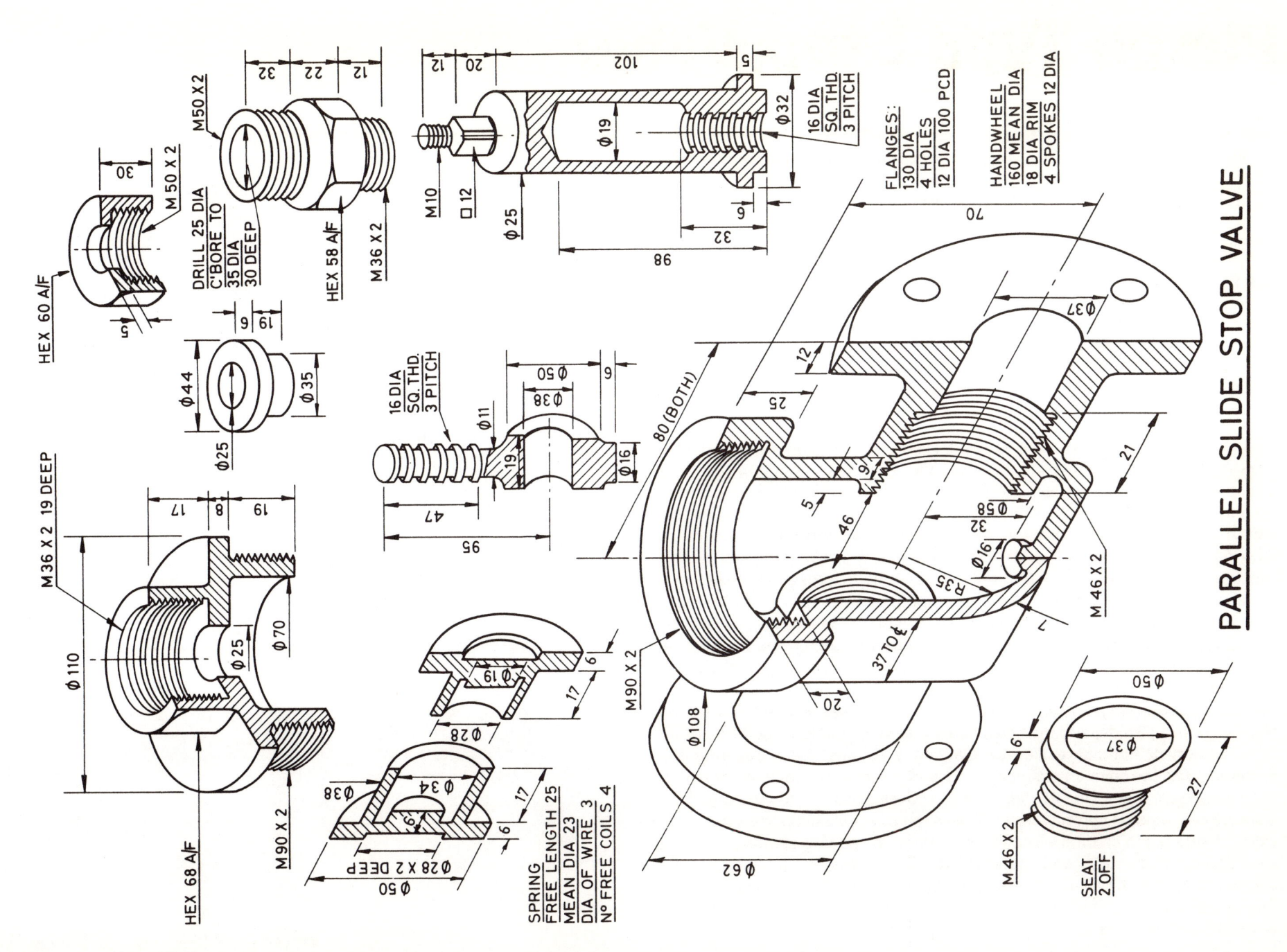

39

Piston Type Stop Valve

DRAW

(i) Front elevation in section through branches showing valve assembled.
(ii) End elevation.
(iii) Sectional plan view through the ports in the lantern.

INCLUDE ON THE DRAWING A LIST OF MATERIALS.

The stop valve shown pictorially on the opposite page is typical of a type used for both high and low pressures. It has the advantage of easy maintenance, in that the renewal of both compressed asbestos fibre rings is all that is required in this respect.

Having decided upon a suitable scale, draw horizontal and vertical centre lines for the three views required, and endeavour to bring along the front elevation and end elevation simultaneously. In end elevation draw concentric circles to represent the 2 BSP thread. (The outer broken circle will be approximately 57 dia. for this thread). On the same centre draw a circle of 82.5 dia. for the spigot. Next draw in a light circle of 90 dia., and around this construct a hexagon using the 60° set square. Project across to the sectional elevation to locate the heights of the hexagonal flanges, which are 25 thick. The tapped holes in these inlet and outlet branches could also be drawn in by projecting from the circles in end view. Continue to build up the body of the valve in section from the dimensions given. The bottom packing ring, lantern bush and top packing ring are now drawn in, and the cover located on top of the upper packing ring. The width of the flange and cover will be the diagonal of a square of 112 side, and thus could be found by a simple calculation using the Theorem of Pythagoras, or by drawing a square and measuring the diagonal. With the cover in position, the top of the piston should be drawn in flush with the bottom of the 45 dia. recess in cover, and then the bottom of the piston drawn in at its correct length. This, of course, will show the valve in its closed position. Finish off the sectional elevation by adding spindle, split nut, cover studs and nuts, and handwheel. The cover, nuts, spindle and handwheel will also be seen in end elevation as well as the bosses, cast on the body of the valve to take the cover studs. Note the diagonal stiffeners 38 wide x 16 deep on cover.

The sectional plan view is required through the lantern ports, and, of course, this could not be attempted until the lantern is located in the sectional elevation. Some imagination is needed here, in order to visualise what remains when the valve is cut through this plane. The solution on page 120 includes this view, together with an outside plan view, which might be required instead of the sectional plan.

For the completed drawing refer to page 120

STOP VALVE (PISTON TYPE)

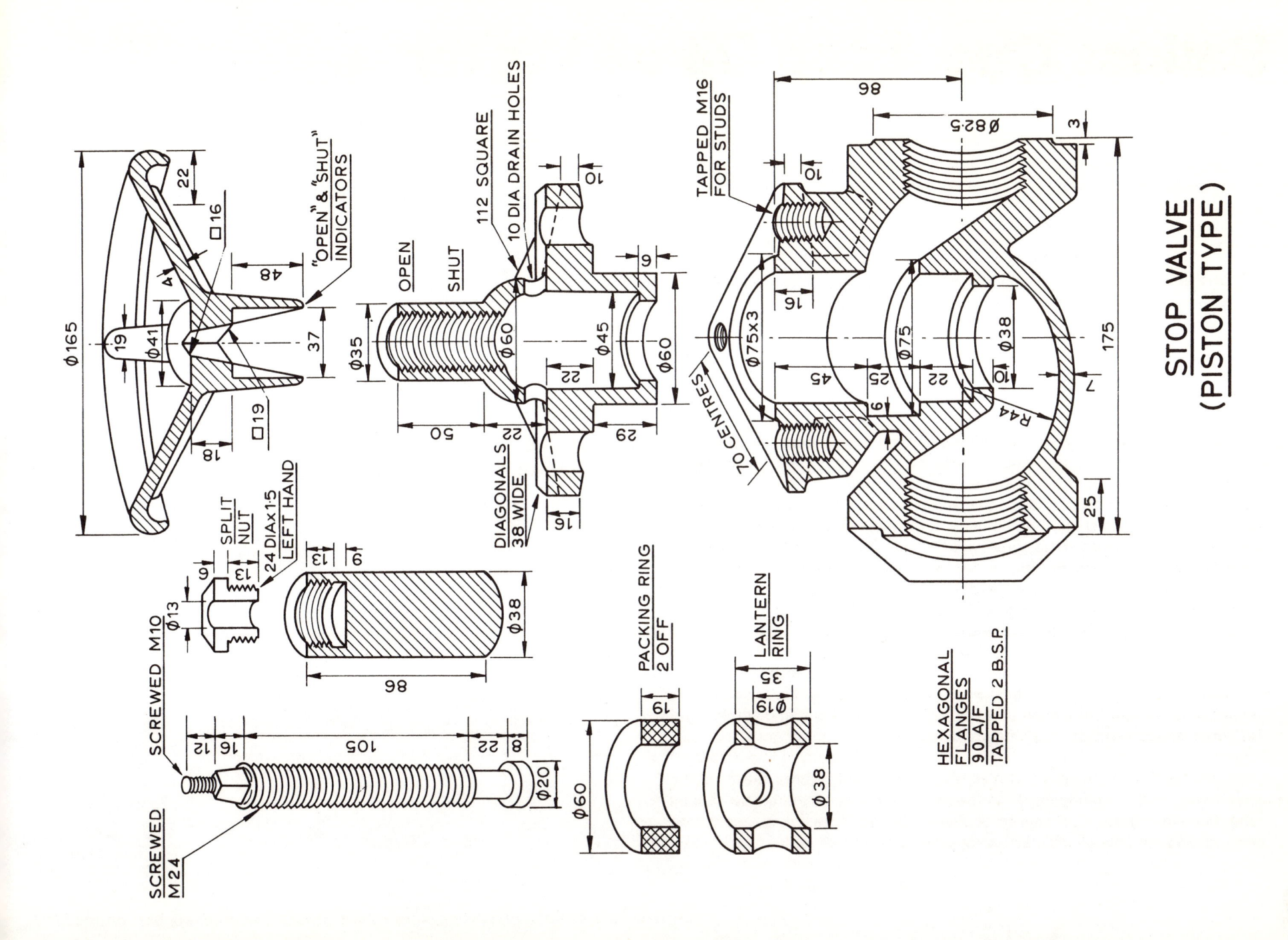

Ballast Chest for Oil or Water

DRAW

(i) Front elevation in section through branches and valve showing all parts assembled.
(ii) Plan view with either dome or blank removed.
(iii) End Elevation.

INCLUDE ON THE DRAWING A LIST OF MATERIALS.

The various components comprising a ballast chest for oil or water are illustrated on the opposite page. The centre branch, in which a screw down valve is fitted, can be common to either the right or left hand branch, depending on which leg the dome is fitted, the dome and the blank being interchangeable.

Commence the front elevation by drawing vertical centre lines for the valve and the two branches, and produce these into the plan view underneath. Draw also a vertical centre line for the end elevation. Set compasses to a radius of 96, and in plan view draw two semi-circles to represent the ends of the chest. This radius could then be used in front elevation, and end view, to mark the ends and width of the chest respectively. The base of the chest could now be drawn in front and end elevations, and used as a datum line for vertical dimensions. Continue with front elevation in section, and draw in the chest from the dimensions given, adding the blank and dome, also in section. Locate the valve in the seat, and draw in the square threaded spindle, these last two items being drawn full. The valve cover and gland are now added to this view, which is finished off with the operating handle. A round-headed screw and washer not shown, could be added to secure the handle. The various heights required in the end elevation are marked off by projecting across from the front elevation, and this view finished off.

In the plan view show the boss with an M24 tapped hole, when, say the blank is removed, this boss being joined to the body by a web 19 thick. The flanges comprising the base will, of course, appear as circles in this view, and where they intersect they should be married together with a small radius.

The completed drawing together with a list of materials appears on page 121

BALLAST VALVE CHEST FOR OIL OR WATER

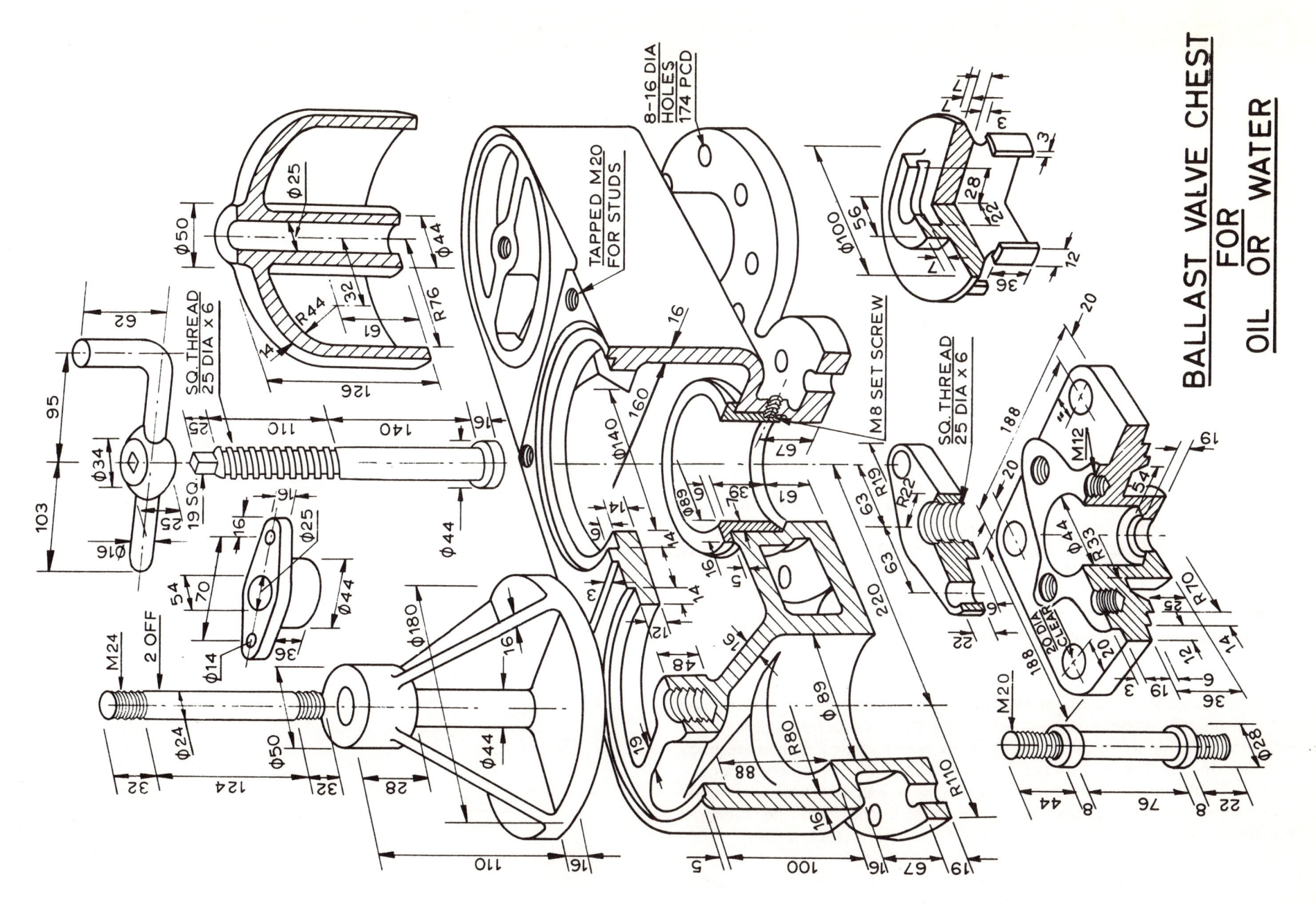

Crosshead and Guide Shoe

DRAW

(i) Sectional elevation of the assembled crosshead and guide shoe, the section plane being taken through the centre of the piston rod, with guide shoe on the left.
(ii) End elevation.
(iii) Plan view with piston rod removed.

INCLUDE ON THE DRAWING A LIST OF MATERIALS.

Present day trends appear to favour the medium speed trunk type engine as a main propulsion unit in a motor ship, but there are still many of the slow speed direct drive engines in service, using a crosshead and guide shoe similar to that illustrated pictorially on the opposite page. Since the first motor ship was built, this type of machinery had few rivals in the Diesel field, and it remains to be seen whether the medium speed gear drive arrangement will supersede slow speed engines.

The design illustrated represents the practice of a particular engine builder, and one of its main features is the fitting of two spherical washers, one above and one below the crosshead. The piston rod cap nut is tightened hard against the end of the piston rod stud, leaving a clearance of 0.1mm on the faces of the spherical washers, to allow for any slight misalignment. The guide shoe is secured to the crosshead by four M36 set bolts in 38 dia. holes, and relative movement is prevented by a stop 95 wide x 40 thick, bolted to the guide shoe and bearing on the top face of the crosshead.

Lubrication of the ahead and astern faces of the guide is effected by oil holes, drilled in the crosshead and guide shoe as shown, the oil entering from the top end bearing. When the crosshead and guide shoe are bolted together, the vertical oil hole in the crosshead comes in line with a vertical hole in the guide shoe, which, in turn, meets a horizontal hole, supplying oil to the ahead face of the guide. At the ends of the oil gutterways in the ahead face, holes are drilled to the astern face for its lubrication. The spherical washers are also lubricated from the oil hole in the crosshead, clearly shown in the pictorial view.

For view (i), draw in the vertical centre line of the piston rod, and around this show part of the piston rod, and the piston rod stud. The top spherical washer could now be drawn in, and as this locates on its flat side in a recess in the top of the crosshead, the latter is now located and drawn in section. The bottom spherical washer, piston rod nut, and locking set bolt finish off the piston rod attachment to the crosshead. Sufficient data is given to locate the guide shoe on the crosshead, and this is now added in section. It should be remembered here that webs are not shown in section.

The end elevation for part (ii) of the question is simply a view looking on the ahead face of the guide shoe, whilst the plan view required for part (iii) should not prove difficult, as top views are given of both crosshead and guide shoe.

For the completed drawing refer to page 122

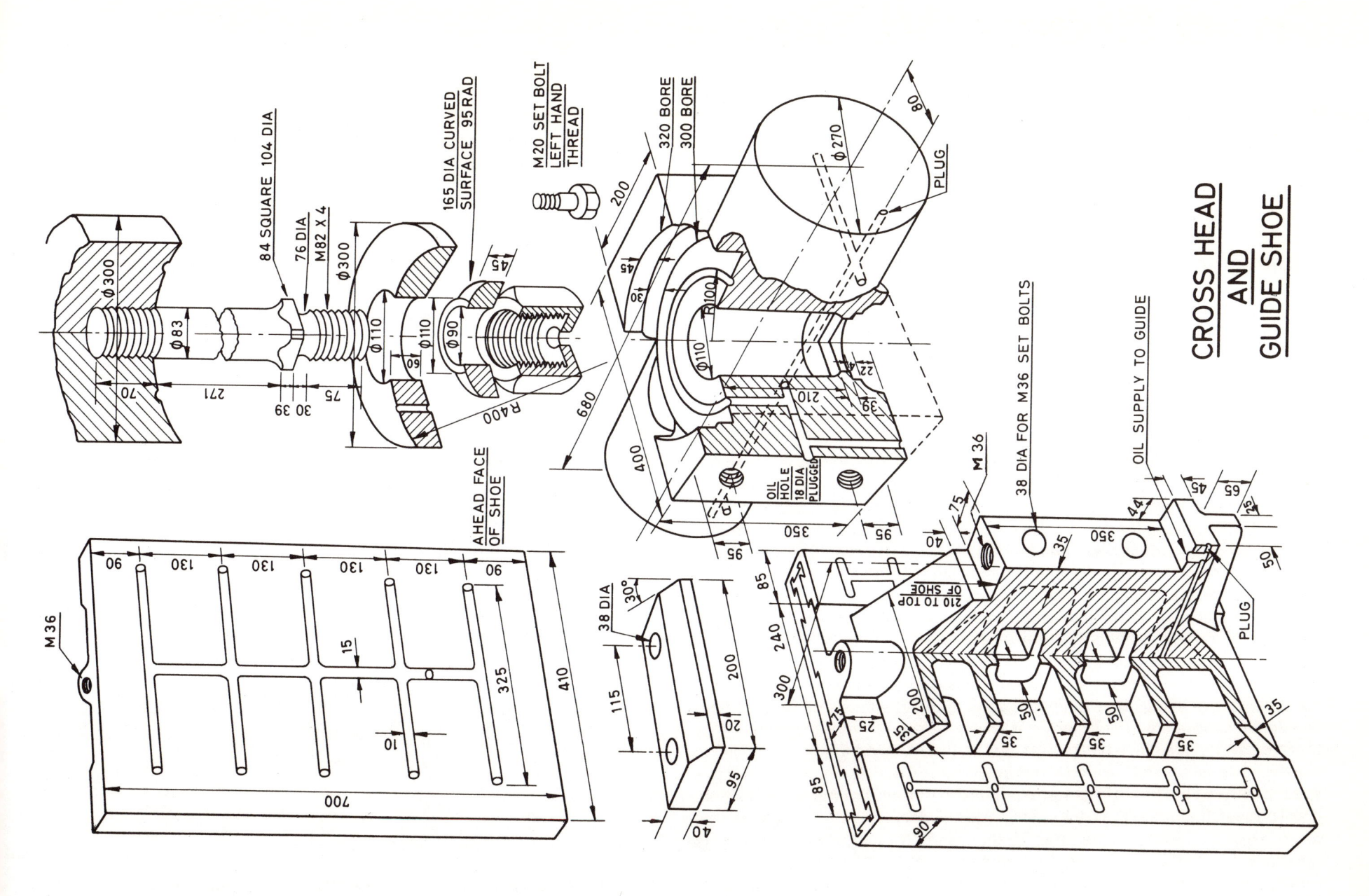

CROSS HEAD AND GUIDE SHOE

Feed Check Valve

DRAW

(i) Sectional elevation of assembled valve with inlet branch on the right, the section plane being taken through centre of valve.
(ii) End elevation.

INCLUDE ON THE DRAWING A LIST OF MATERIALS.

In the Feed Check valve illustrated opposite, the piston formation of the valve eliminates hammering and pulsating, common to ordinary valves. The valve only opens an amount sufficient to pass the quantity discharged. The rocking lever gear for remote operation eliminates the low mechanical efficiency of the spur gear arrangement.

In the sectional elevation, draw in the vertical centre line of the valve and the vertical centre line of the operating spindle 254 apart. Horizontal centre lines for the inlet and outlet branches are now drawn in 200 apart, and the valve body is built up around these from the dimensions given. Locate the valve seat in the screwed part of the body, and then the valve in the seat. Before drawing in the valve spindle, the cover and the operating spindle should be shown. Next locate the operating nut, which is tapped 36 dia. x 4 LH. This could be drawn at the bottom of the screwed portion of the operating spindle, which will correspond to the open position for the valve spindle. A rocker lever is now drawn in, one end locating on the operating spindle nut, and the other end on a collar on a 74 dia. cylindrical nut screwed on the valve spindle. This nut is capable of being adjusted on the valve spindle, and when in the correct position it is locked by a M30 lock nut. It can be seen from the pictorial view that the cover of the valve has guides cast on it, and these are bored 74 dia. to suit the valve spindle nut, which slides up and down in these guides. The valve spindle can now be drawn in to correspond with the valve in the open position. It will be seen that there is a neck bush in the bottom of the stuffing box so that when excessive wear occurs, this bush, rather than the whole cover, is renewed. The plane of the required section in part (i) of the question cuts through the 25 thick web on the cover, and a similar web on the body supporting the boss for the operating spindle, but being webs these are not shown in section.

The various heights in the end view are obtained by projecting from the sectional elevation, and in this view two rocker levers will be seen end on, and 57 apart, with the operating spindle nut sandwiched between them.

On the completed drawing on page 123, a plan view of the valve is also included, but in view of the time allowed for the question, this view would not be required

FEED CHECK VALVE

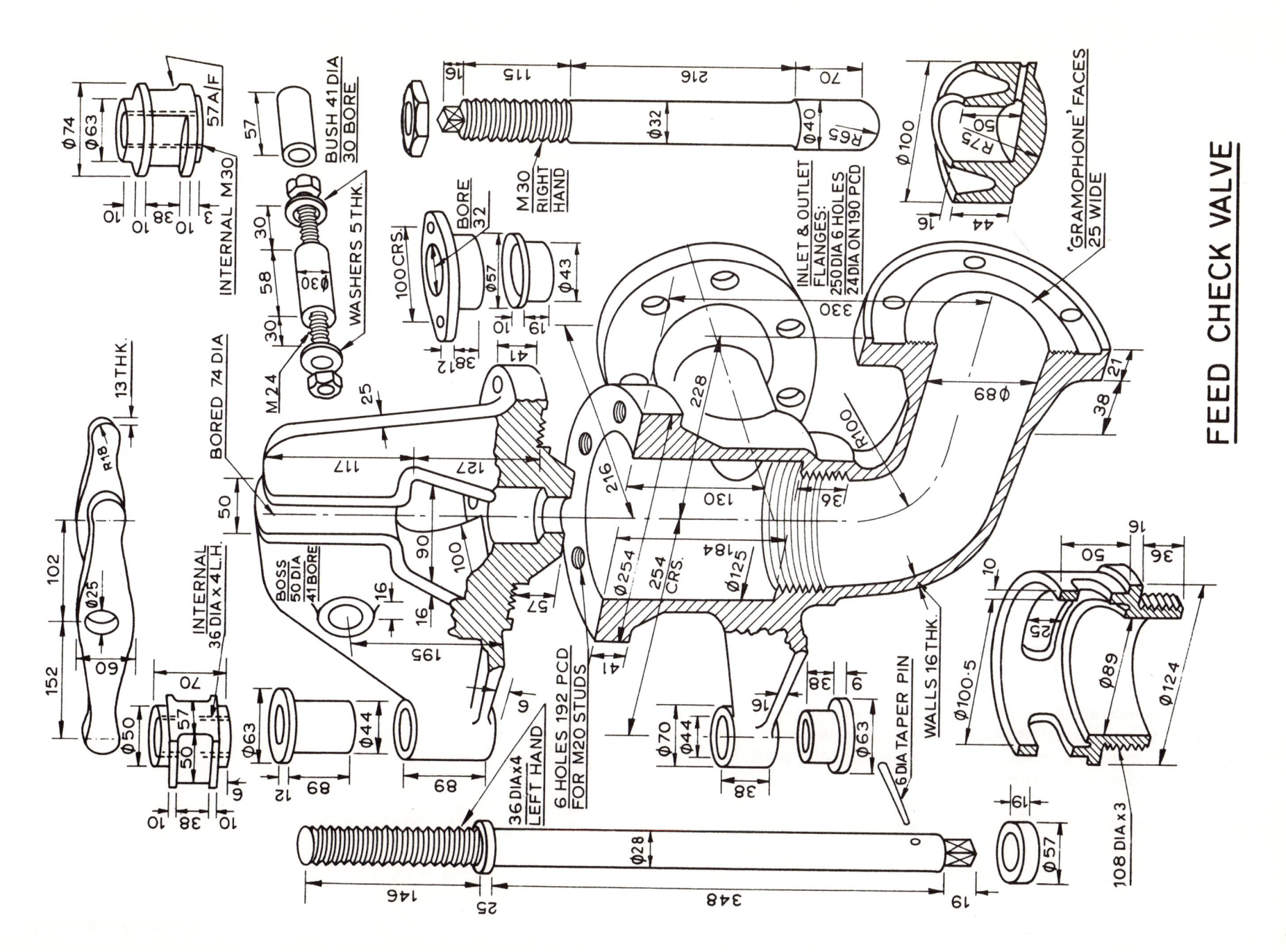

47

Gear Pump

DRAW

(i) Sectional elevation through longitudinal axis of driving shaft showing pump assembled.
(ii) End view, one half an outside view looking on gland with coupling removed, the other half a section through pump casing, looking on the ends of the pinions.

INCLUDE ON THE DRAWING A LIST OF MATERIALS.

Gear pumps generally come under the heading of constant displacement pumps, and as such do not require priming. The increasing, decreasing volume of the suction and discharge sides of the pump is provided by the action of the gears coming out of mesh, and meshing. It is important, therefore, that the direction of rotation of the pump is correct in relation to the suction and discharge ports. Gear pumps are best suited for duties on the more viscous fluids, and for that reason they are very common as lubricating oil pumps.

Commence the sectional elevation by drawing in the horizontal centre line of the driving shaft, and produce to the end view. Note that this shaft is 37 dia. in way of the pinion, and 36 dia. for the remainder of its length. The driving pinion is located 30 from the right hand end of the shaft, and the centre line of the driven pinion drawn in at 84 above the centre line of the driver. The pinions, in this view and also in the end view, are illustrated in the manner shown by the convention on page 11, and students are also advised to refer to page 17, where the terms used in involute gearing are explained. Sufficient detail is given in the pictorial view for the pump body, end cover, gland and coupling to be drawn, and further detail of the ports is given in a local section, so that these can be shown in the end view. Note that there is a clearance of 0.1 between the outside diameter of the gears and pump casing, but this, of course, is too small to show on the drawing.

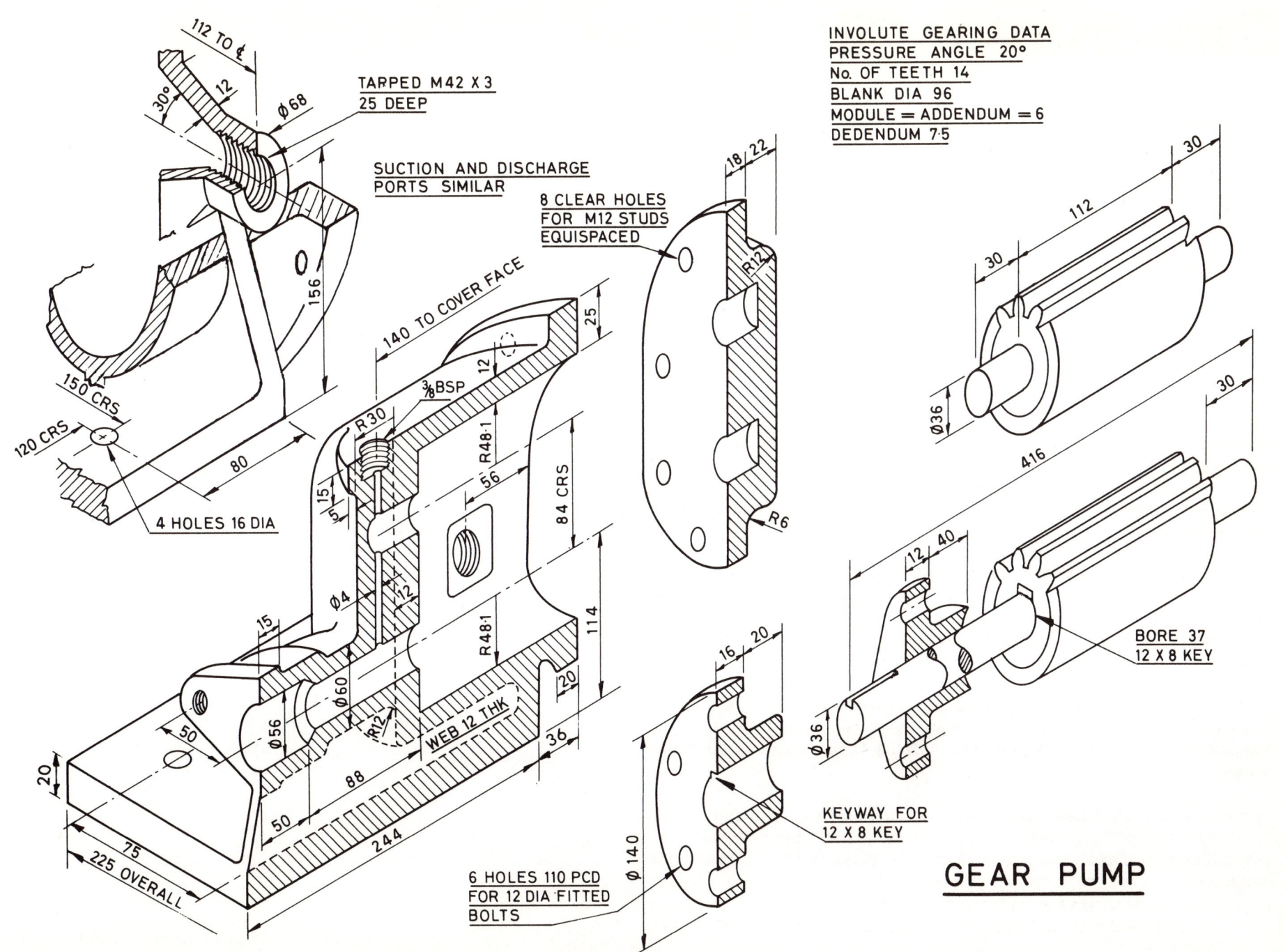

GEAR PUMP

Starting Air Pilot Valve

DRAW

(i) Front elevation in section showing valve assembled.
(ii) End view.
(iii) Plan view, one half to be in section through atmospheric port, other half a section through port to automatic valve.

INCLUDE ON THE DRAWING A LIST OF MATERIALS.

The pilot valve illustrated pictorially on the opposite page is operated by the engine starting lever, and functions to relieve the air pressure on top of the piston in the automatic valve (see page 126). When the starting lever is moved from the 'Stop' to 'Start' position, the spindle of the pilot valve is moved up and the 20 dia. cylindrical portion of the valve plugs the 20 dia. hole in the body, thus shutting off the air from the reservoir, and permitting the line to the automatic valve to vent to atmosphere. In this position the automatic valve is open to the starting valves. When the starting lever is moved further to the 'Fuel' position, the pilot valve spindle drops, thus shutting off the atmospheric port, and allowing the line to the automatic valve to pressurise and hence shut the automatic valve.

The assembly of this valve is comparatively easy as there are only two components to locate, and even if a student had not seen one previously, common sense would indicate that the 45° mitre on the valve locates on a similar mitre forming the valve seat machined in the valve body. The ¾ BSP plug obviously fits in a similar hole tapped in the top of the valve body.

Commence the sectional elevation by drawing in the vertical centre line, and build up the valve body around this from the dimensions given, then, as explained above, locate the valve spindle and plug. These two items will not be shown in section.

In the end view, the bosses for the inlet and atmospheric connections will appear as circles of 45 dia. and 64 dia. respectively, with tapped holes in each boss indicated by the conventional method. In this view the 19 dia. hole at the bottom of the atmospheric port will appear as an ellipse, the minor axis of which is found by projecting from the sectional elevation, whilst the major axis will be 19, and it should be noted that part of the spindle will be seen through this hole. One method of constructing an ellipse is given on page 12.

The plan view will show on the left of the centre line, a section through the 1¼ BSP port, and on the right, a section through the 1 BSP port. In this view also, the 19 dia. port at an angle of 15° to the horizontal will be seen as an ellipse. Finish off by showing that part of the fixing flange which is visible, together with the 12 dia. clear holes for bolting down.

For the completed drawing refer to page 125

STARTING AIR PILOT VALVE

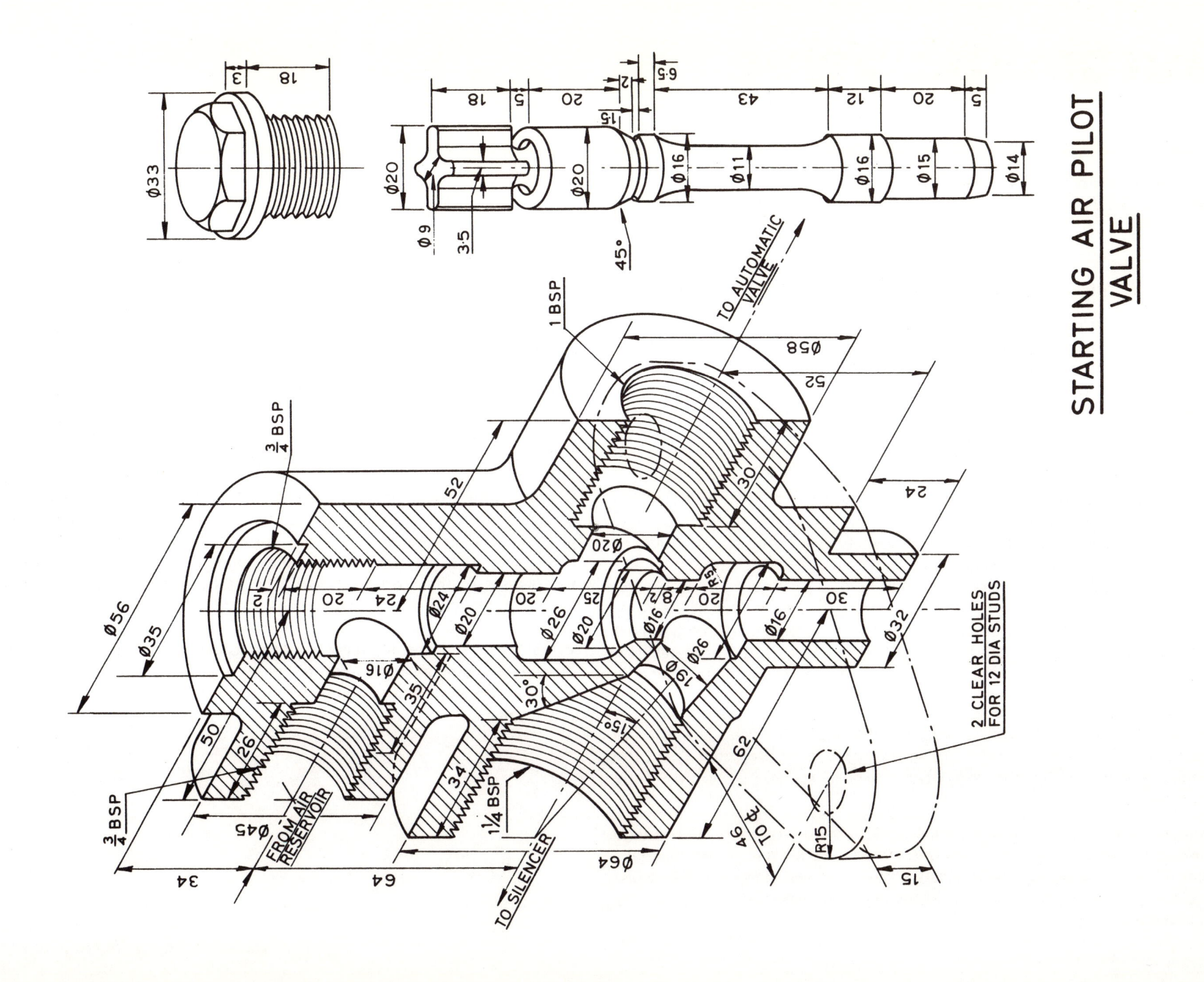

Automatic Valve

DRAW

(i) Elevation in section with all parts assembled.
(ii) A plan view, the bottom half to be a section through engine port.

INCLUDE ON THE DRAWING A LIST OF MATERIALS.

From studying the pictorial view of this drawing, it will be seen that the valve body is a rectangular block 335 square by 403 long, with a central bore of various diameters, and three ports drilled into this bore, two on one face and one on the opposite face. This block should be drawn at the outset, in section as the question requires. The liner or bush, to take the piston on the top valve, could now be drawn, and the top cover located on this by **its spigot**. It will be noticed that the spigot on the top cover is 1mm longer than its recess in the flange of the liner, so that the joint is made by the spigot, leaving the cover 1mm clear of the valve block. The valve cage could now be located, the bottom diameter of 152 fitting into the recess in the bottom of the valve block. This recess being 12 deep, will allow the valve cage to be 8 proud of the bottom of the block, and as the recess in the bottom cover is 7 deep, the joint will be made in the recess, leaving the bottom cover 1 clear of the valve block. The top valve is next located on its seat, which is machined round the top of the valve cage, and the bottom valve drawn in directly underneath this, both valves being held together by the spindle and a M39 nut, locked by a split pin. The spring is located on top of the piston and in the 110 dia. recess in top cover.

The plan view could then be drawn in, and in the bottom half which is to be a section through the left hand, or engine port, the valve cage is shown in section, with its ports 20 wide. These ports are now projected up to the elevation, and made 70 high in way of engine port, and 80 high in way of the atmospheric port.

As this valve is only used when manoeuvring the engine, there will be long periods when it is not in use. In order that the valve may be occasionally turned in its seat, a hexagon is provided at the lower end of the spindle, whilst a special tool is located under this hexagon to exercise the valve in a longitudinal direction.

OPERATION OF AUTOMATIC VALVE
(referring to page 126)

Operation of the engine starting lever moves a double seated pilot valve and cuts off the air supply to upper space of the automatic valve above the piston, and vents it to atmosphere. As the piston area is greater than the upper valve area, the air pressure from the reservoir pushes the valve assembly up against its spring, opening the top valve, and shutting the bottom valve. From the automatic valve, starting air now flows through the starting line to the spaces under the pistons of the starting air valves on the cylinders. Air also passes to the starting air distributor, forcing the distributor pistons on to the cams, and so, as the cams permit, allow pilot air to pass to the spaces above the pistons of the starting valves. The starting valves open in sequence, permitting the engine to turn in the desired direction.

When the engine starting lever is moved to the fuel position, the pilot valve is opened, allowing air to be admitted to the top of the automatic valve piston, thereby closing it, and opening the starting line to atmosphere through the lower valve.

For the completed drawing refer to page 126

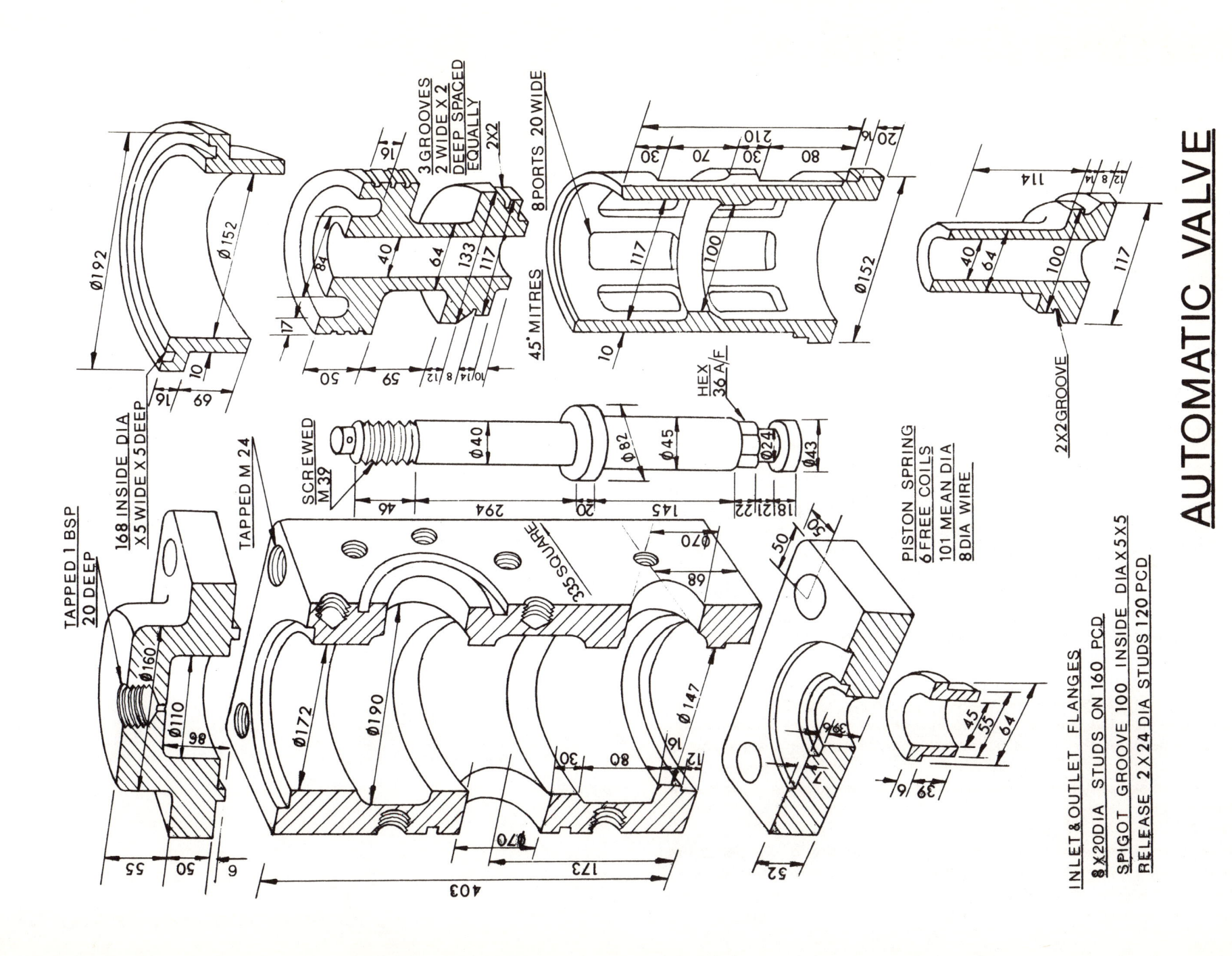

AUTOMATIC VALVE

Starting Air Valve

DRAW

(i) Sectional elevation through inlet branch showing all parts assembled.
(ii) End elevation.
(iii) Sectional plan through vent in valve body.

ADD A LIST OF MATERIALS.

The various components comprising a starting valve are shown pictorially on the opposite page. From his experience of this type or a similar type, the student would be required to know how the valve is assembled. Starting air is admitted to the valve through the inlet branch, and into a space between the 35 dia. piston machined on the spindle, and the valve. As piston and valve are of the same diameter, there is no tendency for the air to open the valve. At the correct time in relation to the position of the engine crank, air from the starting air distributor enters the ⅜ BSP connection in the valve cover. This air acts on the 75 dia. piston which is fixed to the spindle, and forces the valve open against the compression of a spring fitted below the piston. When the distributor has turned through the required angle, air is cut off from the space above the 75 dia. piston, and the spring pressure closes the valve.

Draw in vertical and horizontal centre lines for the three views required, the horizontal centre line for the elevations being drawn through the inlet branch. In order to find the total width of the body, the equilateral triangle representing the body flange, and the inlet branch should be drawn in plan view. Project up from this view to locate the body in elevation. Continue with sectional elevation by adding the valve seat. This item is screwed 58 dia. x 2.5 at one end, to fit into a similar tapped hole in the body. The valve spindle can now be located on the seat, and drawn in together with the 75 dia. piston, sandwiched between the 27 dia. collar and an M20 nut. Show valve spindle and nut full, and piston in section. Continue with this view, adding cover, guide bush and nut, and the spring which is located in the groove in the bottom of 75 dia. piston.

Provision is made in the valve body for a liner in which the 75 dia. piston fits, and there is also a liner for the 35 dia. piston, secured by a M6 grub screw drilled half in the liner, and half in the valve body. In the end view the heights are projected across from the sectional elevation, whilst diameters and widths generally for this view are taken from the other two views.

In the plan view required, the 80 dia. part of the body will be seen in section, with a 25 wide port cut in it. The section plane will also cut the spindle which is 15 dia. at this point.

For the completed drawing refer to page 127

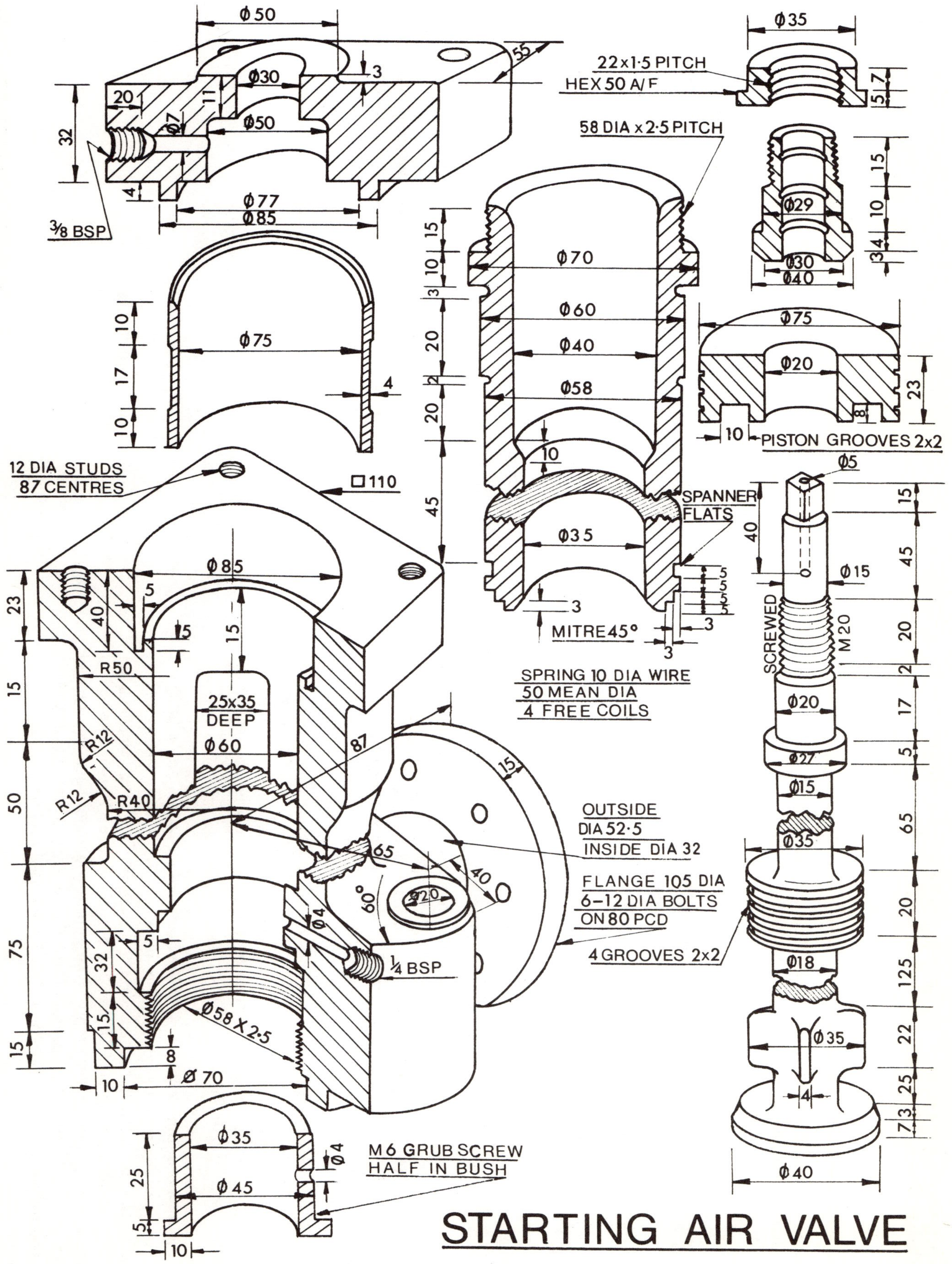

STARTING AIR VALVE

Burner Carrier

DRAW

(i) Sectional elevation showing burner carrier assembled.
(ii) Sectional end elevation through fuel inlet cock.
(iii) Plan view.

INCLUDE ON THE DRAWING A LIST OF MATERIALS

Since harbour authorities generally throughout the world tightened their regulations with respect to air pollution, the burner for which the carrier illustrated on the opposite page was designed, has gone out of favour. Steam assisted atomisation of the fuel requires a more sophisticated burner and carrier, but for many years the type illustrated was very common on board ship.

The burner, of circular cross section, fits into the elongated hole in the carrier, and the M20 jack screw is tightened to make an oil tight joint between a port in the burner and the conical end of the nipple, which is screwed into the oil inlet manifold. This manifold is secured to the burner body by an M12 stud and a set screw, the thimble end of which protrudes into the elongated hole, to locate with a hole in the burner. This arrangement, together with a 12 wide slot in the carrier, to engage with a key on the burner, ensures that the burner is inserted in the correct position, and that the inlet port in the burner body locates on the conical end of the fuel nipple, before the jack screw is tightened. As a precaution against withdrawing the burner when the fuel cock is in the 'On' position, the handwheel on the jack screw is covered by the operating handle of the cock.

Draw in the horizontal centre line for the sectional elevation, and produce to the end view. Draw in the body of the burner, noting that the entry end is elongated by 16, and the furnace end has a bore of 40. The flange securing the carrier to the furnace front is 132 dia., and one of the three 13 dia. holes could be shown on the vertical centre line. Mark off the holes for securing the fuel inlet manifold, and the clear hole for the nipple. Locate the manifold in section and add the nipple, also in section. The 12 dia. stud and thimble ended set screw are shown full.

The sectional end view through the inlet cock will show the bracket cast on the top of the carrier, to take the operating spindle of the cock. On the vertical centre line of the cock, which is 65 from the centre line of the inlet nipple, draw in the tapered plug cock in the open position. As no instruction is given as to whether the cock should be open or closed, the plan view is probably easier if the cock is drawn in the open position. In this view the gland, neck ring and nipple will appear in section, but the cock and jack screw will be shown full. Finish off this view by drawing in the handwheel in section, and the cock operating handle covering this item.

In the plan view, draw in the cover on the cock operating handle, and the portion of the handwheel which can be seen. Part of the gland will be seen, together with the gland nuts, the nut securing the cock handle to the spindle, and the nut securing the handwheel on the jack screw.

For the completed drawing refer to page 128

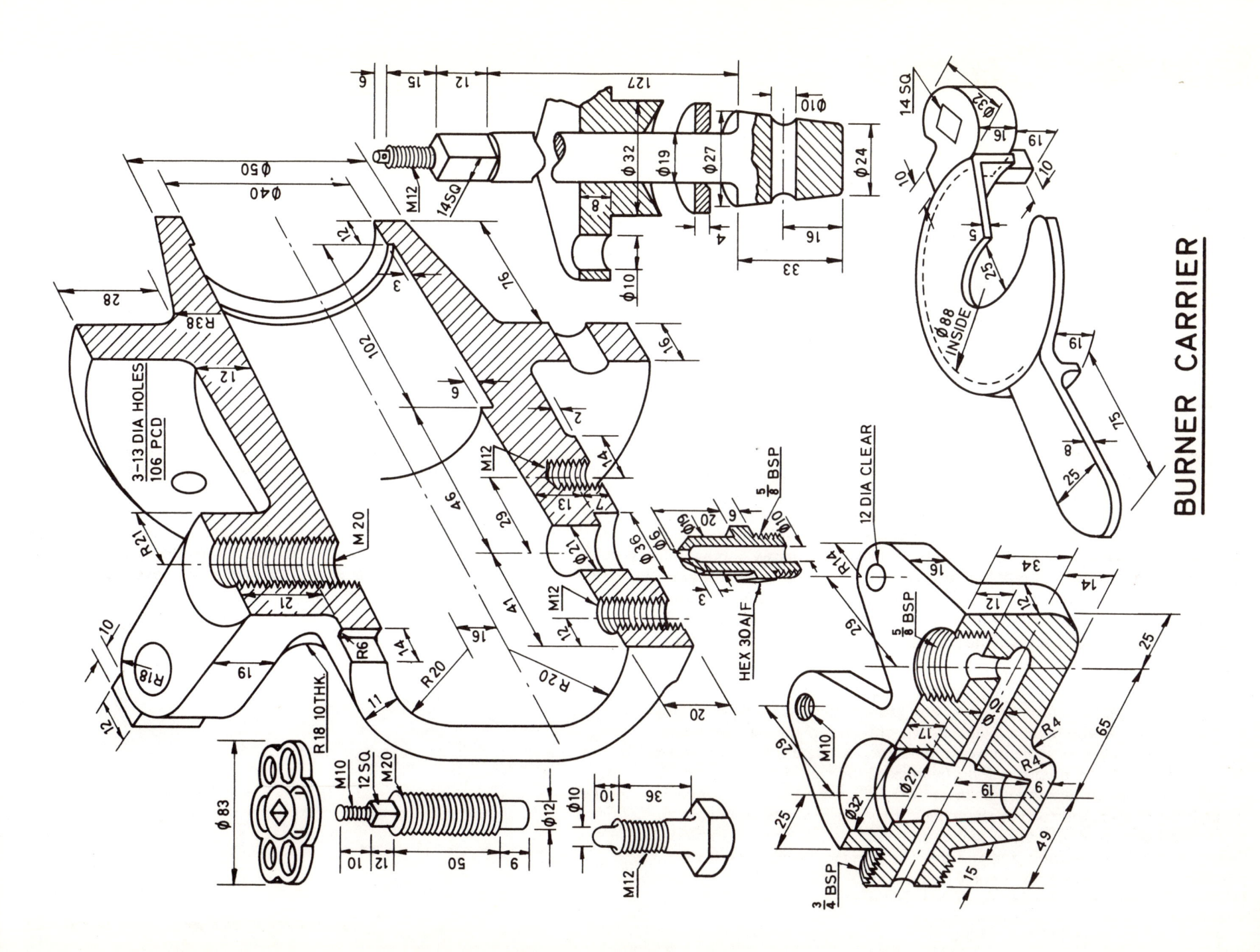

BURNER CARRIER

Burner Carrier

DRAW

(i) Front elevation showing carrier assembled. This view is to be partly in section through fuel handle locking trigger.
(ii) End elevation.
(iii) Plan view.

INCLUDE ON DRAWING A LIST OF MATERIALS.

The main features to be observed in the burner carrier illustrated opposite are the locking device for the fuel handle, which prevents fuel being turned on until a burner is in its working position, together with a safeguard against withdrawal of a burner whilst the fuel is on.

A key or lug is welded on the burner, and this locates between the two 15 dia. collars on the spring loaded locking trigger. As the burner is tightened against the fuel inlet nipple, the trigger is moved to the left, thereby disengaging the 10 dia. pin from a hole in the fuel handle, and allowing the latter to be turned to the 'On' position. To prevent withdrawal of the burner whilst fuel is on, a cam is formed on the fuel handle which prevents the jack screw handle from locating on the square portion of the jacking screw to unscrew it. Only when the fuel handle is in the 'Off' position, is there sufficient clearance between fuel handle and jacking screw handle to allow the latter to engage on the square. The fuel shut off cock is of the parallel plug type, with sleeve packing.

Commence the front elevation by drawing horizontal centre lines for the fuel cock, locking trigger and jacking screw, and around these draw in the profile of the carrier from the details given. When drawing in the locking trigger details, assume a burner to be in working position in the carrier, so that the right hand end of the trigger will be flush with the right hand side of the carrier. These details will of course be shown full, as the required section is through the trigger. The section plane could finish slightly above this detail, and the top portion of the carrier shown full, with details of the fuel nipple, (hidden by guide plates), and jacking screw, shown by broken lines. Both the fuel handle and jacking screw handle could now be drawn in, and the view finished off by showing the two guideplates, the various plugs, and the fuel inlet coupling at 57 from the vertical centre line.

When the profile of the end view has been drawn in project across for the centres of the three hexagons seen in this view, and finish off by showing the fuel inlet coupling, the top guide plate and the fixing lug, with an M12 tapped hole. Two more M12 holes would be required in the back of the carrier for fixing purposes, but these have been omitted for clarity.

A plan view projected from the front elevation would finish the drawing, and in this view it would be in order to omit the two handles.

On the completed drawing on page 129, an elevation looking on the operating end is shown, but in consideration of the time allotted for the question this would not be required. This view should be studied however, as it may be asked for in place of one of the other views.

For the completed drawing refer to page 129

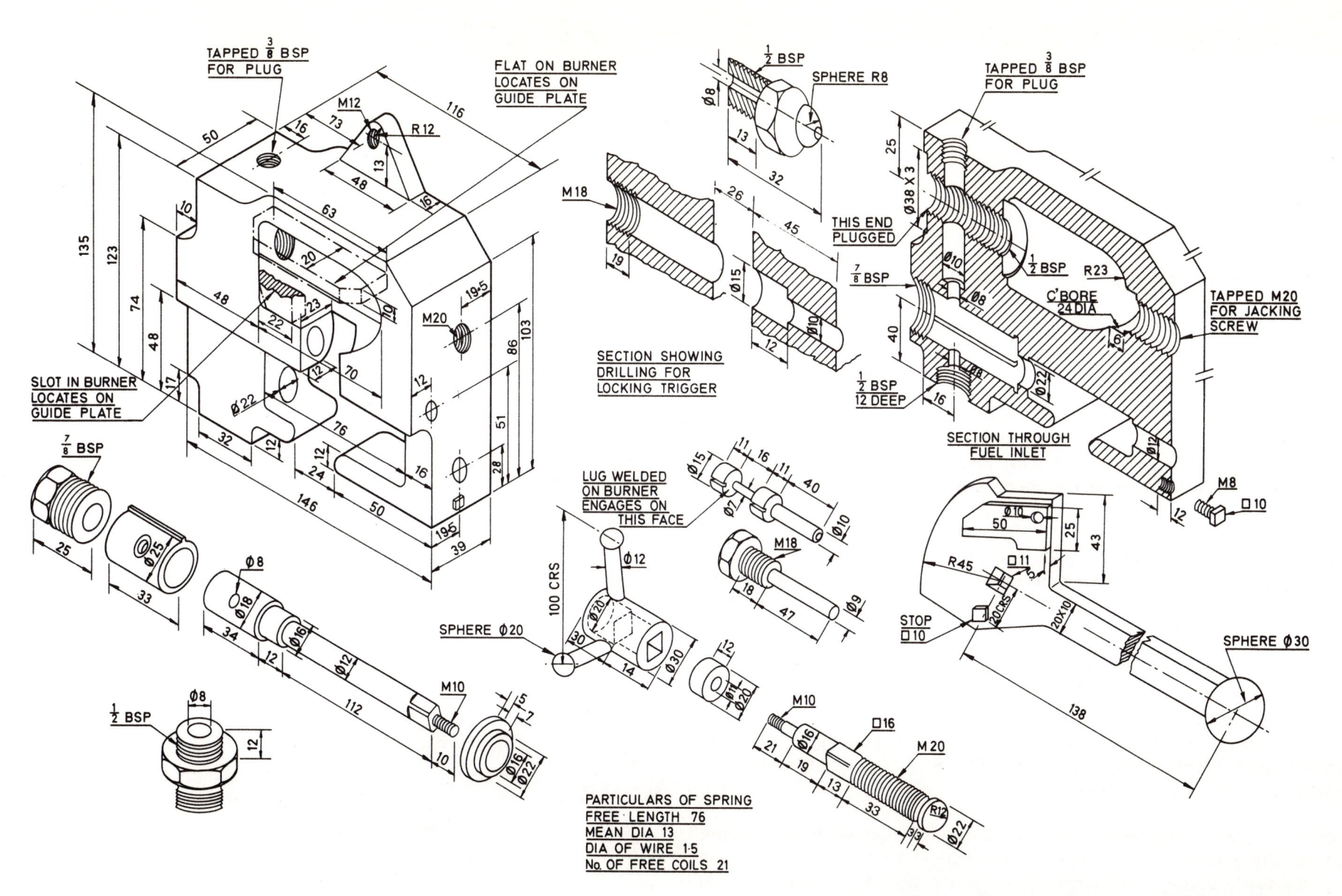

OIL FUEL BURNER CARRIER

Connecting Rod and Bearings

DRAW

(i) Front elevation, one half in section through axis of bottom end bearing, and one top end bearing in section through axis of crosshead pin, showing all parts assembled.

(ii) End view showing bottom end bearing, half in section through bottom end bolt, and top end bearing, half in section through top end bolt.

INCLUDE ON THE DRAWING A LIST OF MATERIALS.

The pictorial view of a large connecting rod with its top and bottom end bearings is given on the opposite page. The design of this assembly differs slightly in detail from one engine builder to another, but basically there is not a great deal of difference. It will be seen that the connecting rod is 250 dia., and at the top and bottom ends of the rod there are flats 250 thick machined on it. Where these flats meet the fillet radius of the palm at the bottom, and the forked end at the top, curves will be seen, and the projection of these is an important feature of the drawing. For the construction of these curves refer to page 16.

The longitudinal centre lines for both views should be drawn first, then the centre lines of the top and bottom end bearings. The rod should be broken, so that a reasonably large scale can be used to show bearing details distinctly. Sufficient detail is given in the pictorial view, for the bottom end bearing to be formed on the end of the palm, noting that there is a spigot of 120 dia. by 25 deep in the top half, which fits in the central oil hole in the connecting rod. Only one top end bearing is to be shown in section, and in this view show a similar spigot 80 dia. by 20 deep in the bottom half of the top end bearing, which fits into a recess machined in the forked end. These spigots reduce the shearing stress in the top and bottom end bolts. Show the whitemetal shells dovetailed into the cast steel bushes.

In the end view, one bottom end bolt and one top end bolt will be seen in the half sections required, and the locking arrangement for the nuts on these bolts will also be shown. Notice that liners of a total thickness of 6 are fitted in the bottom end bearing, and similar liners of thickness 4 are fitted in the top end bearings.

For completed drawing refer to page 130

CONNECTING ROD AND BEARINGS

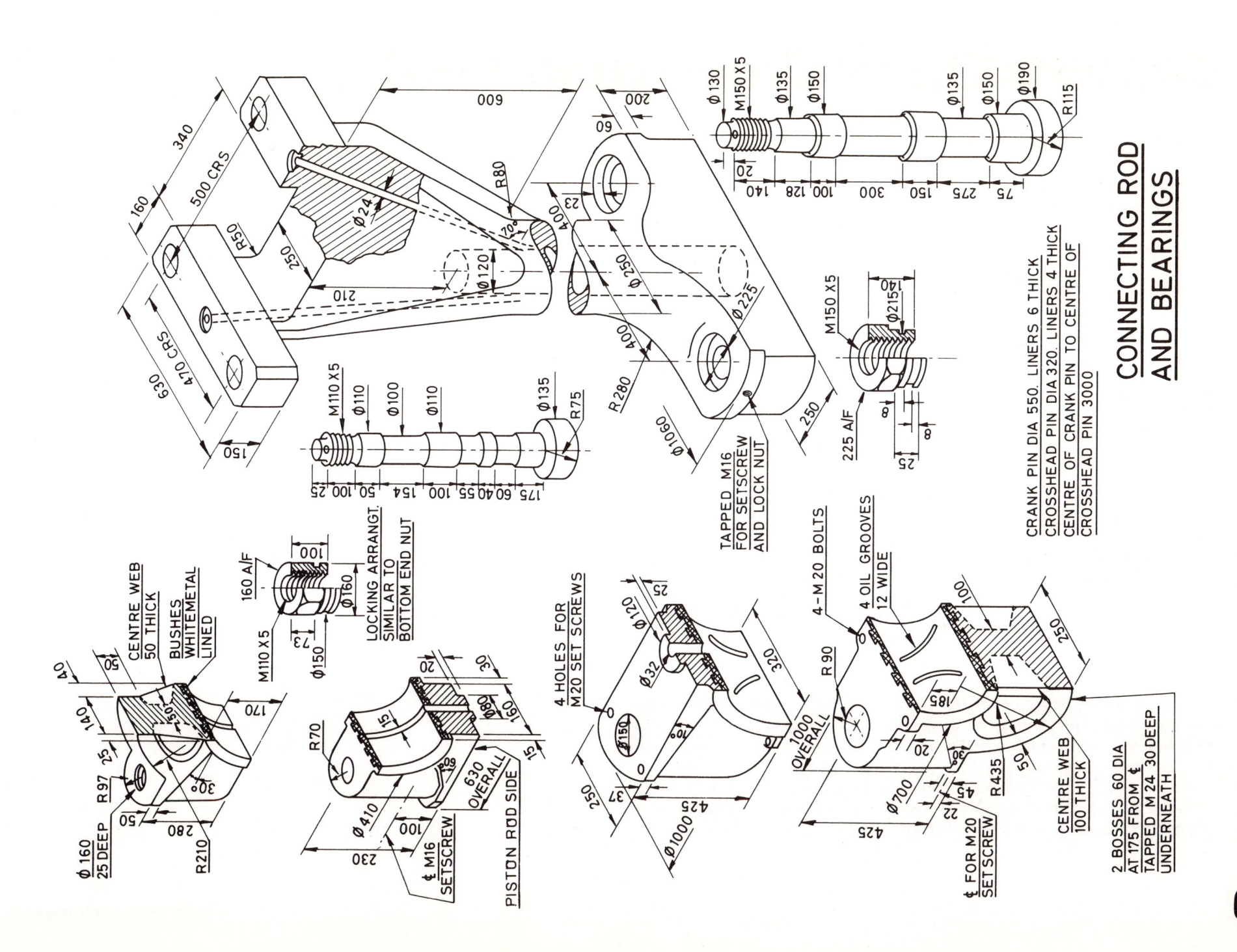

Gravity Davit Centrifugal Brake

DRAW THE FOLLOWING VIEWS:—

(i) Sectional elevation showing all parts assembled.
(ii) End view with the handwheel and end cover removed.

APPEND A LIST OF MATERIALS.

Gravity davits irrespective of the means of hoisting the lifeboat, must be fitted with a hand brake for deliberate regulation, and a centrifugal brake for the automatic control of the lowering speed between the limits of 18m and 30m per minute. Because it is out of sight and used so infrequently, engineers tend to forget its existence, and in many cases the centrifugal brake does not receive the attention it requires for the safety of the occupants of the lifeboat. Details of such a centrifugal brake are given in the pictorial view.

From the overall diameter of 584 decide on a suitable scale, and draw in horizontal centre lines for both views. Build up the elevation around this centre line by drawing in the 86 dia. shaft and hub, showing two of the 20 dia. pegs, one at the top and the other at the bottom. It is quite in order to turn the brake shoes half a pitch in this view, in order to show more clearly the springs and anchor pins which otherwise would be hidden by the pegs. The centre line of the springs can be calculated from the dimensions given, and these, together with two shoes, can be drawn in. The brake drum end plate and hand wheel will finish this view.

The shape of the brake drum in end view is obvious from the pictorial view, and the circles representing the various diameters could now be drawn in. Note that the twelve stiffening ribs are not parallel to the axis, so that more than the semi-circles of their cross-section will be seen in this view. Next draw in the PCD of the guide pegs, and draw these in with the brake shoes in between them. A clearance of 1.5 is given between each shoe. The hub boss, springs, and square on end of shaft to take crank, would finish this view.

For solution refer to page 131

GRAVITY DAVIT
CENTRIFUGAL BRAKE

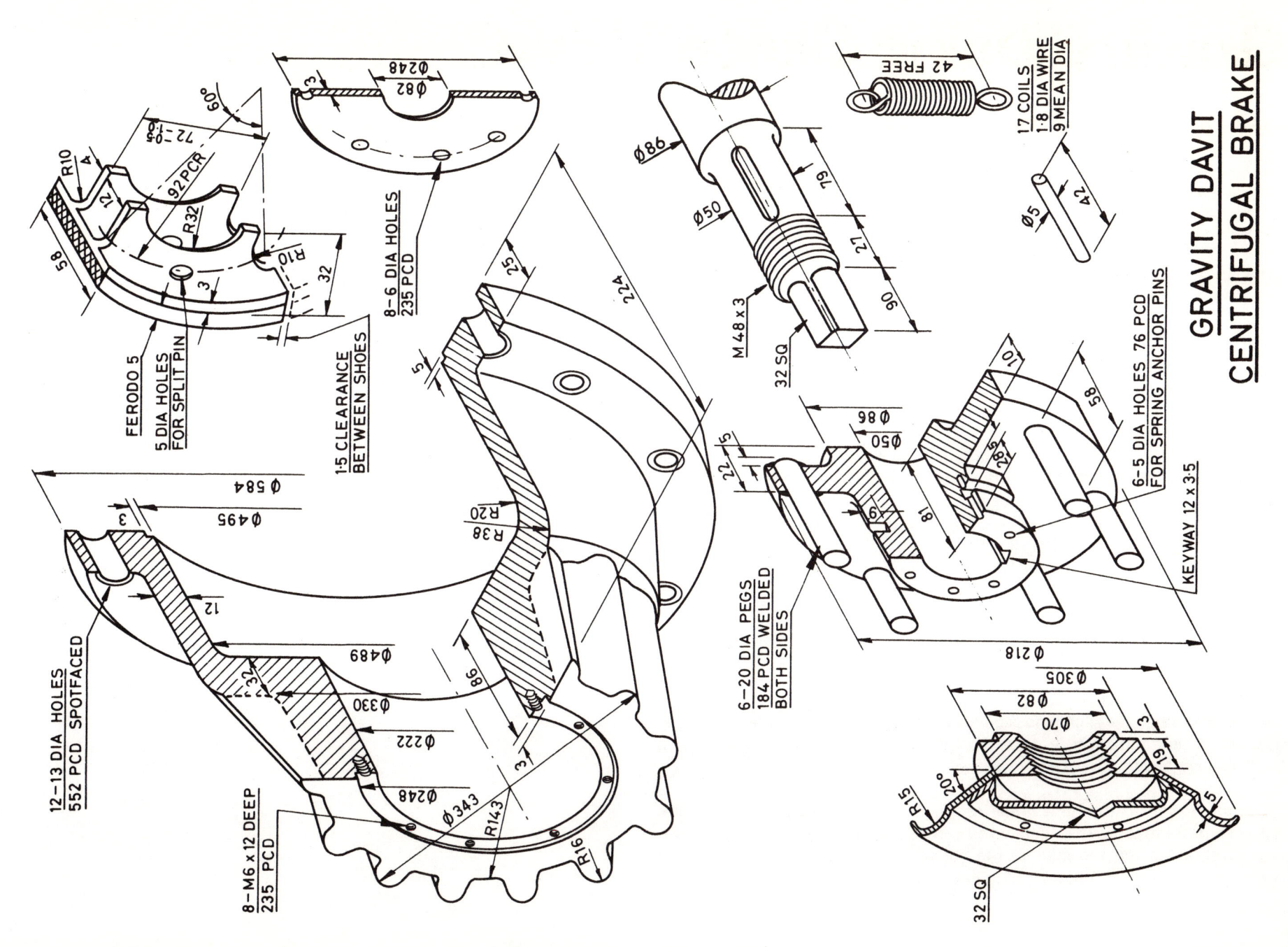

63

Quick Closing Sluice Valve

DRAW THE FOLLOWING VIEWS

(i) Front elevation in section showing valve assembled, and link gear upright.
(ii) End view in section. The link gear in both the above views need not be shown in section.
(iii) Plan view with handwheel removed.

APPEND A LIST OF MATERIALS.

Most marine engineers will be familiar with the operation of the quick closing sluice valve, shown in the pictorial view opposite. The link gear is clearly visible, whilst the internal details are similar to an ordinary sluice valve, with the addition of a heavy spring to collapse the links in an emergency, and hence close the valve.

With the link gear in the upright position, the valve can be operated as an ordinary sluice valve, but should the valve be required to be shut quickly, a sharp pull on a wire rope, led to a position remote from the valve, will collapse the links and achieve this. The valve end of the wire rope is spliced to the bridle, fitted at the junction of the links.

Draw vertical centre lines for both elevations, and a common horizontal centre line through the bore of the valve. Using these as datum lines, build up the valve body, valve wedge and spring, all in section. It is advisable to draw valve in the shut position. The spindle and the top cover are now drawn in, and with the hinge pins on the cover as a datum line the link gear is then drawn in. Note the stops on the cover to allow the bottom links to locate slightly past the vertical, in order that a slight knock does not collapse the links. A tabernacle is also cast on the top cover, to take the boss of the bridle, when the valve is being used as an ordinary sluice valve.

The plan view with the handwheel removed is finally drawn, most of the dimensions for this view being taken by compasses from the other two views. A view showing the trip gear in the collapsed position is included in the solution, as this may be asked for in the examination.

For solution refer to page 132

QUICK CLOSING SLUICE VALVE

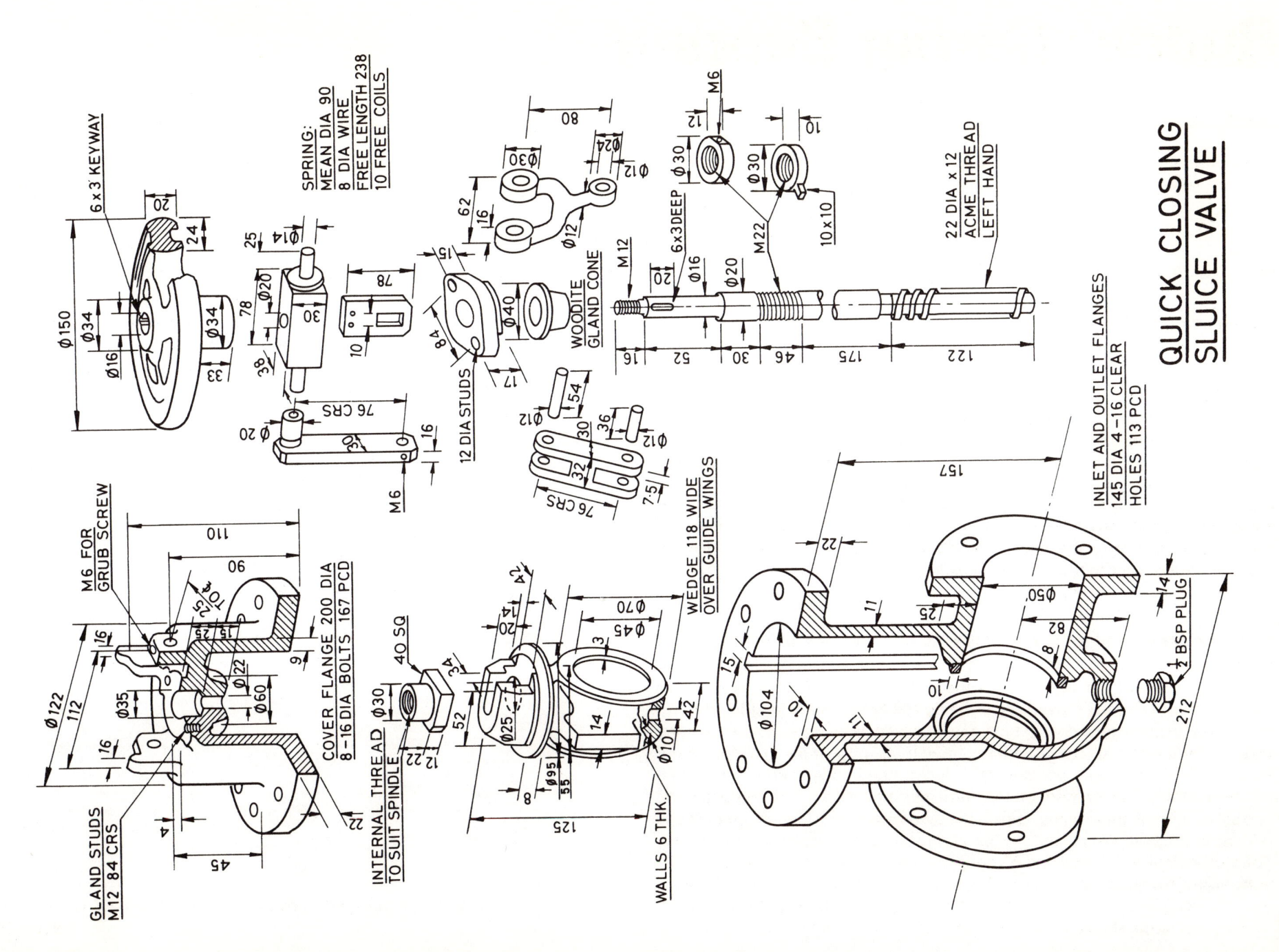

Rudder Carrier Bearing

DRAW

(i) Elevation, one half in section through a lubricator hole, other half an outside view.
(ii) Plan view.

INCLUDE ON THE DRAWING A LIST OF MATERIALS.

A rudder carrier or thrust bearing is shown pictorially on the opposite page. This bearing is out of sight on many ships and so is inclined to be overlooked by junior engineers.

Basically it consists of a moving cast iron cone keyed to the rudder stock, and a fixed cast iron cone bolted firmly to the deck, usually in the steering flat. Cones and gland ring are made in halves so that they can be fitted over the stock when in place, and are held together by fitted bolts. Four grease lubricators are fitted to lubricate the bearing faces, and gutterways in the top cone, ensure lubrication across the whole bearing surface. The thrust on the upper cone is transferred to the tiller by a distance piece sandwiched between these. An alternative method, is to fit a split collar in a recess round the stock, flush with the top of the upper cone.

Draw the vertical centre line of the stock and produce to the plan view underneath. A 204 dia. circle in the plan view, and two vertical lines in the elevation represent the stock. Locate the bottom cone or base around this in elevation, from the sizes given. By bringing the two views along together, it will be found that a good deal of time can be saved by using a compass setting in both views. Although the lugs for the fitted bolts in the moving cone, are not in line with the lubricator holes through which the half section is required, it will be in order to show them on the same cutting plane in the elevation, the correct relative position being seen in the plan view. Next draw in the top cone with its key in both views, and the gland ring could then be fitted around this. The stuffing box is 30 wide and 63 deep, which allows for two turns of 30 dia. soft packing to retain the grease in the bearing. The oil gutterways 1.5 deep in way of the lubricator hole, and the 6 dia. hole for lubricating the cylindrical part of the bearing should be shown in the sectional elevation.

Finish off the plan view, by showing the fitted bolts holding together both cones and the gland ring. Two long 30 dia. studs going through bosses of 37 radius cast on the base flange, and also in the hollow part below the fixed cone, prevent this cone from spreading.

The solution to this question, together with a list of materials used, appears on page 133

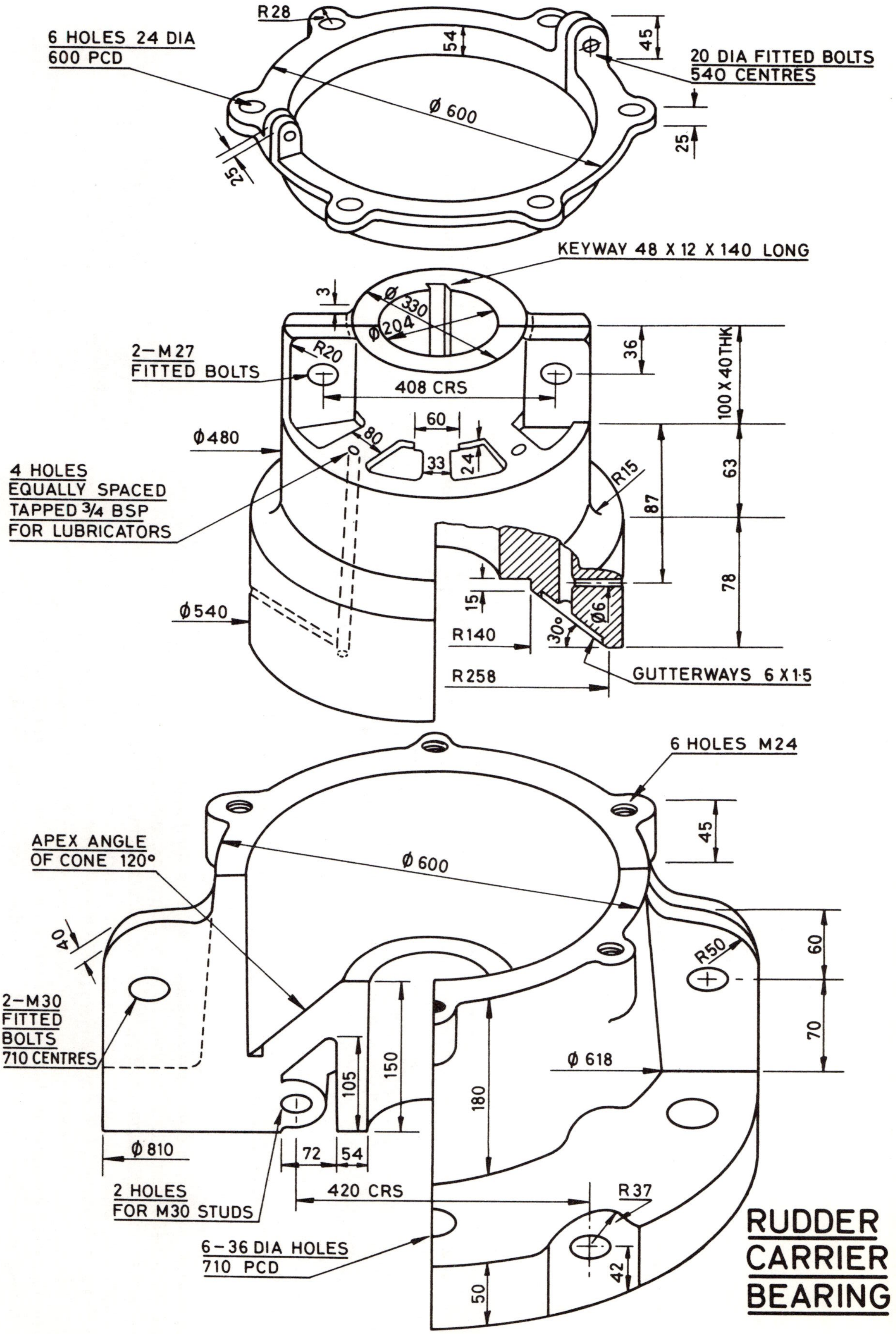

RUDDER CARRIER BEARING

Reducing Valve

DRAW

(i) Elevation in section through centre of valve showing all parts assembled, and with inlet branch on the left.
(ii) End elevation.
(iii) Plan view, one half in section just below boss on top cover, other half an outside view.

INCLUDE ON THE DRAWING A LIST OF MATERIALS.

High pressure steam enters at the top left hand connection, is throttled on its passage through the valve, and leaves at a lower pressure through the right hand branch. At the bottom of the casing there is a rubber diaphragm, sandwiched between two discs secured to the valve spindle, and the steam pressure acting on this, forces it down against the compression of a spring. The effective areas of the diaphragm and valve are such, that the pressure tending to open the valve is balanced by the downward force on the diaphragm. Additional pressure tending to close the valve, is obtained from the reduced steam pressure acting on top of the valve, and this is balanced by the upward force of the spring. The spring may be adjusted to give varying reduced steam pressures, and pressure in excess of the spring setting will tend to close the valve, causing further throttling. Water from the condensation of the steam in the lower part of the casing, protects the rubber diaphragm from the direct heat of the steam. A pressure gauge connection is provided on the reduced side of the valve, and a relief valve is also fitted on this line to protect it, should the reducing valve fail to function.

In the pictorial view of the valve on the opposite page, the various components have been illustrated in their correct relative positions, so that no difficulty should be experienced in locating these in the orthographic projection. In the required sectional elevation, build up the valve body around a vertical and horizontal centre line through the branches from the dimensions given. The valve and valve seat are shown to a bigger scale in the pictorial view, and from the details given these items are now located. The spindle can now be drawn in, the dome on the top end of which is 1.5 clear of the end of the valve. Locate the 8 thick diaphragm on the bottom face of the body, and the various items on the screwed portion of the spindle above and below the diaphragm, could now be shown. Locate the bottom cover over the diaphragm and draw this in, together with two 12 dia. bolts securing it to the valve body. Finish off this view by showing the two 22 dia. pillars, spring holders and spring, and the spring adjusting set screw in the bottom flange.

The end view projected from the front elevation, will show the overall width of the body as 214, and details of the four webs cast on the valve body, will also be seen in this view. The plan view will show one half of top cover with three securing nuts, whilst the other half will show the 90 bore portion of the body in section, and half of the 70 dia. top flange of the valve.

For the completed drawing refer to page 134

REDUCING VALVE

Four Stroke Piston & Rod

DRAW THE FOLLOWING VIEWS

(i) Elevation in section showing all parts assembled.
(ii) End elevation, one half in section through a piston rod stud showing securing arrangement, other half an outside view.
(iii) Plan view in section, the section plane to be taken between conical and cylindrical parts of piston.

INCLUDE IN YOUR DRAWING A LIST OF MATERIALS FOR THE MAIN COMPONENTS.

The pictorial view shows details of a piston, and piston cooling arrangement for a four stroke direct drive Diesel engine. Although this type of engine is seldom used now for main propulsion, if one considers a piston to be essentially a piston ring carrier, piston design does not alter a great deal from one type of engine to another.

From the piston dimensions given, decide upon a suitable scale for the size of paper. The length of the piston rod may be ignored for this purpose, as it is usual in drawing long components of uniform section to break them, in order to avoid a long narrow drawing.

Draw first the vertical centre lines for both elevations, and a light line to represent the bottom of the piston in each view. The top of the piston can now be represented by a light horizontal line in each view, before drawing in the crown radii, when the diameters are marked off. Note that the piston is partly cylindrical, and partly conical, this being to allow for extra expansion at the combustion chamber end. It will be seen that two section planes are used in the pictorial view, one view giving details of one of the ten bosses for the piston rod studs, the other a view between these bosses. The sectional elevation required for view (i) could be taken between the bosses, and the details of the piston drawn in from the information given. The rod could now be drawn in, the 560 dia. flange on the rod locating in the 560 dia. recess in the piston. The cooling arrangement, piston rod nut, and locking plate would finish this view.

Complete view (ii) referring to stud details etc.- and use a tee square to project across for piston grooves, and other heights similar in view (i). The piston rod may be broken off in this view, as sufficient detail will be given in view (i) for its manufacture.

The sectional plan view will show mainly the position and shape of the ten bosses, for the 48 dia. studs securing the rod. The six **16 dia.** studs securing the piston cooling assembly should also be shown in this view.

For solution refer to page 135

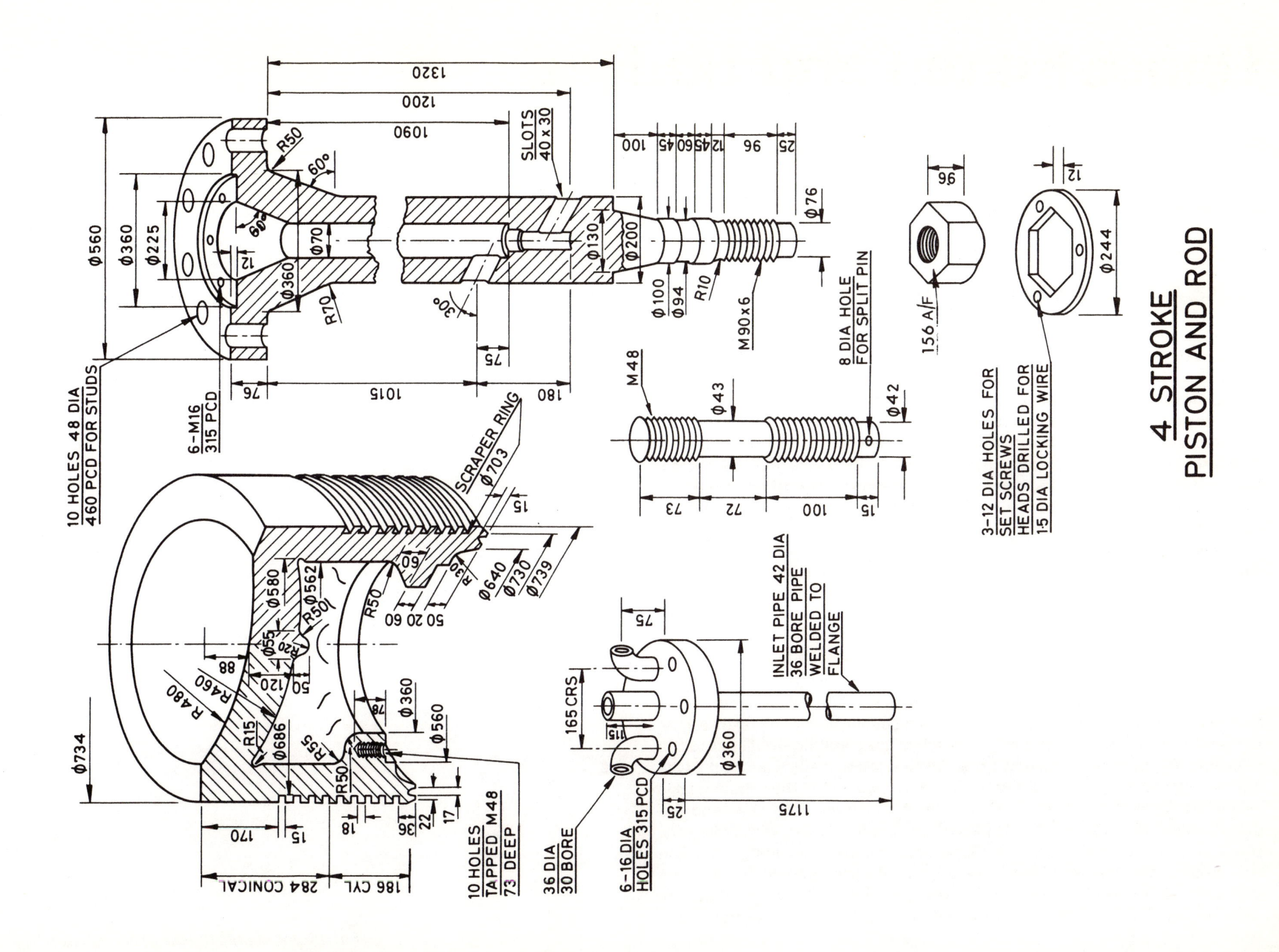

4 STROKE PISTON AND ROD

Upper Piston and Rod

DRAW

(i) Elevation in section through piston cooling inlet and return ports in rod, showing all parts assembled.
(ii) Plan view with skirt cover removed.

INCLUDE A LIST OF MATERIALS ON THE DRAWING.

In a two cycle engine where the exhaust ports in the liner are uncovered by the piston, it is usual to fit a cast iron skirt for this purpose. This skirt, in the case of the piston illustrated opposite, is secured to the rod by twelve M20 studs, whilst the piston head is secured to the rod by eight M33 studs, screwed into 86 dia. bosses cast in the head. With this construction, piston cooling oil is able to flow between the bosses. Rubber rings are fitted in recesses 40 bore x 6 wide around the 33 dia. studs, to seal the space in the skirt from oil leakage. Two rubber rings are also fitted around the top flange of the rod, to retain the piston cooling oil in the piston. The whole assembly is secured to the transverse by two 50 dia. studs, screwed into the end of the rod, and the skirt cover is sandwiched between the transverse and a spigot recess in the end of the rod. This spigot has a dia. of 310 whilst the hole in the cover is 320.

Around the vertical centre line, produced to the plan view, draw in the piston rod in section, showing both piston cooling ports. Locate the skirt on the 470 dia. spigot on the rod, and draw this in, showing two M20 studs securing the skirt to the rod. The 12 thick webs cast in the upper portion of the skirt can be left until they are shown in plan view, when they will be projected to the elevation. The piston head is now drawn in by locating the bosses on the face of the flange on the piston rod. It will be noticed that, when the piston head is located as described, there is a clearance of 3 between the skirt and the piston head (112-109). Two M33 studs could now be shown, with a distance piece on each under the nut. Finish off the piston head by indicating the piston ring grooves, the cast iron bearing ring, and the three grooves for relieving thermal stresses. At the top end of the skirt, the cover could now be drawn in, together with the two M50 studs for securing the piston to the transverse beam.

In the plan view with the cover removed, show the twelve studs for securing the skirt, equidistant between twelve radial webs. Two of these webs will be seen in the sectional elevation, and their location in this view is found by projecting up from the plan view.

For the completed drawing refer to page 136

UPPER PISTON AND ROD

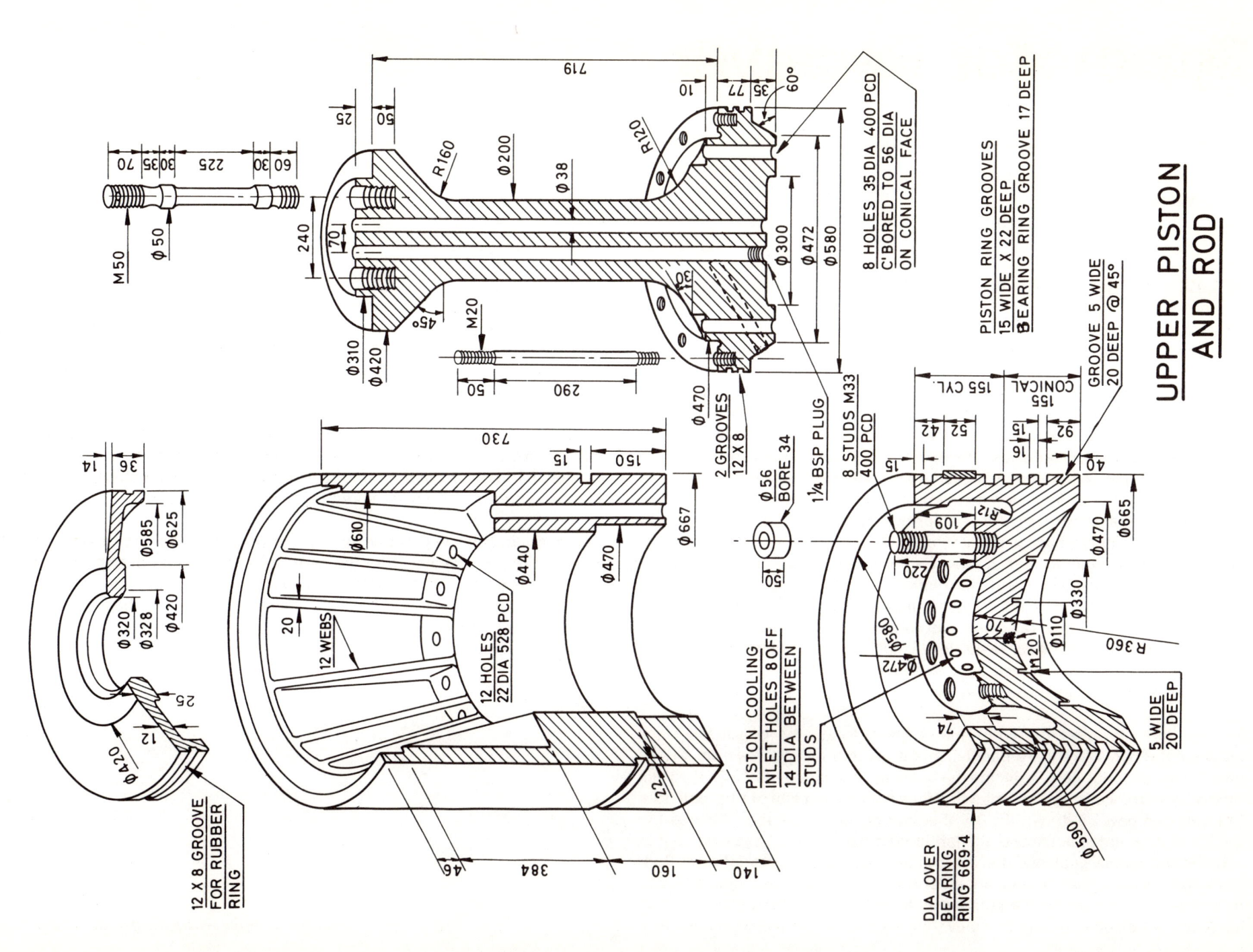

Telemotor Receiver

DRAW

(i) Front elevation of the assembled telemotor.
(ii) End elevation.
(iii) Plan view in section through centre of rams.

INCLUDE ON THE DRAWING A LIST OF MATERIALS.

The two receiver cylinders are bolted together by two 120 square flanges, and between these flanges is sandwiched a crosshead, which by means of a 20 dia. pin, transmits the lateral motion of the cylinders to a 40 dia. rod, connected by links to the variable delivery pump. From the outer end of each ram, a short pipe is led to its circuit valve, and beyond the circuit valve the pipe continues to the top, or bottom, of the transmitter cylinder. The receiver rams are screwed at their outer ends, and fixed to brackets bolted to a common bedplate, and remain stationary, whilst the cylinder body as a whole is free to move along the rams. Any such movement on either side of mid position is made against the compression of two springs, which bring the receiver, and with it the transmitter, back to mid position when the wheel is let go. The 20 dia. pin referred to above can be located in either of two positions. When inserted in the cross-head, the 40 dia. operating rod is moved by the hydraulic system, and when inserted in a screwed sleeve (not shown), the rod is moved by the hand gear. This gear (also omitted) consists of two bevel pinions, one of which is tapped to take the screwed sleeve, and is bolted to a base cast on the left hand bracket, and operated by a handwheel locally, or by an extended spindle arrangement, to a steering wheel aft.

For the front elevation (drawn in mid position) locate the crosshead block between the two 120 square flanges, and draw in the cylinders from details given in the pictorial view. To locate the brackets, one guide rod 1276 long will next be drawn in centrally, and allowing for a nut on each end of this, both ram brackets can now be drawn in. Another nut on the inner face secures the guide rod. As pointed out above, the left hand bracket has a base cast on it to take the emergency gear, and this is fully illustrated in the pictorial view. Both rams can now be drawn in, as a shoulder on each bears on the inside face of its bracket, and the screwed end takes an M48 nut. Part of the gland nuts on each end will also be seen in this view, but most of these will be hidden by a spring compression sleeve over the guide rod. One of the springs which is seen in this view is illustrated in the conventional manner which is shown on page 9. To locate the operating rod, it should be noted that the free end of this rod is flush with the outer face of the bracket.

For the sectional plan view required for part *(iii)*, project down to obtain the relative position of each item. The cylinders and crosshead will be seen in section in this view, but the rams could be shown full. Four stops, 52 dia. x 24 bore, are fitted on the guide rods and secured by Allen screws.

The end elevation will show mainly the profile of the ram bracket which supports the emergency steering gear, and this can be drawn in from the dimensions given in the pictorial view. Although no details are given of the radius joining the guide rod boss and the emergency gear base, this can be drawn in to please the eye.

For the completed drawing refer to page 137

TELEMOTOR RECEIVER

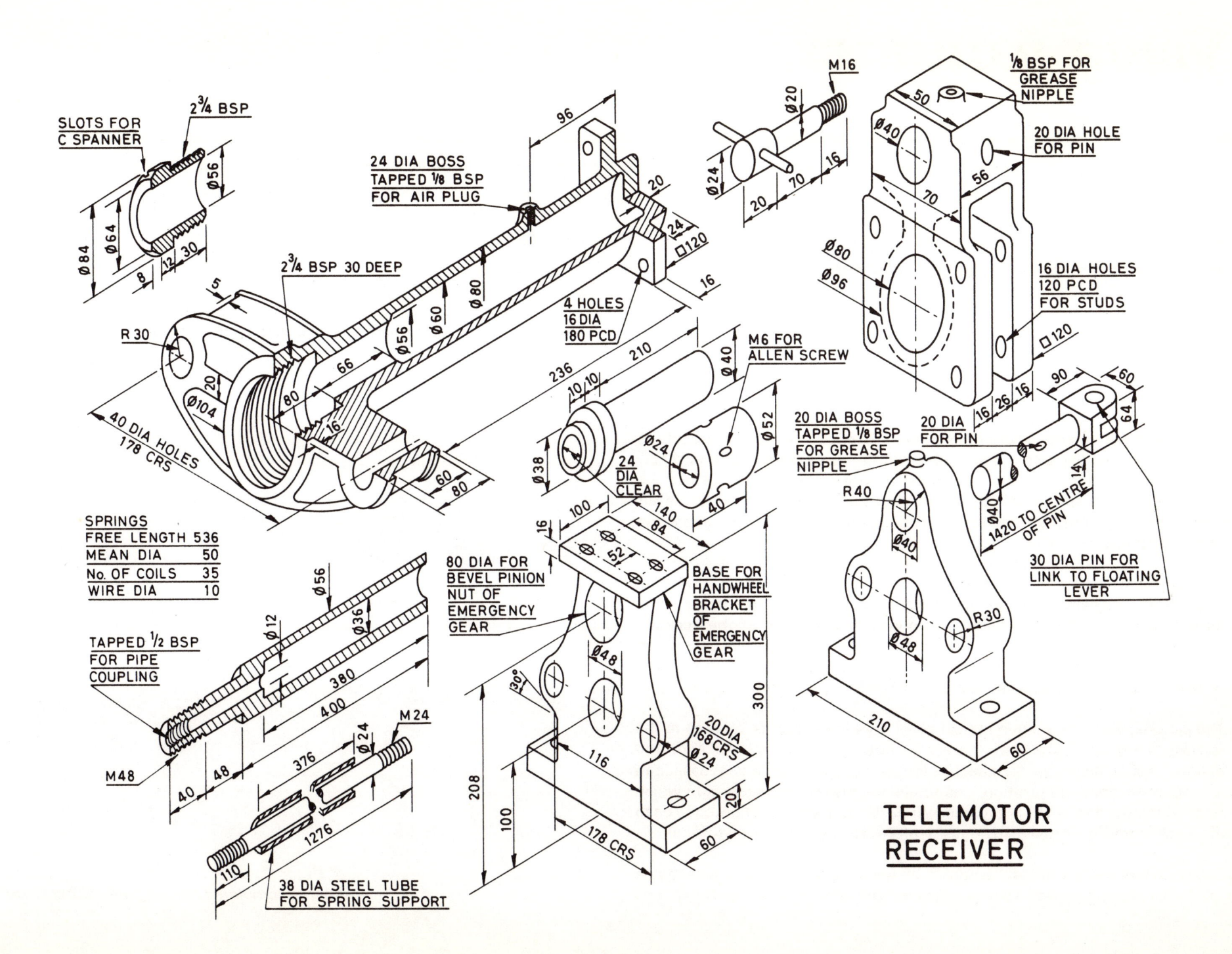

Valve Actuator

DRAW

(i) Elevation in section through the assembled actuator, the section plane being taken through the air inlet connection.
(ii) End elevation.

INCLUDE ON THE DRAWING A LIST OF MATERIALS.

The valve actuator shown pictorially opposite, is a device for positioning a valve plug in relation to its seat, hence regulating the flow in the line to which it is fitted. The valve is not included in this exercise, but it would be secured to the actuator at the boss on the bottom of the yoke, and its spindle would be screwed into the tapped hole in the end of the actuator spindle.

Referring to the assembled drawing on page 138, air pressure on top of the diaphragm forces the piston down, and with it the spindle, which is rigidly attached to the piston. This downward movement is made against the compression of a spring, which returns the piston, on a reduction of air pressure. The spindle therefore takes up a position in direct relationship to the control air pressure.

Should the air pressure fail, or for any other reason the actuator become inoperative, the piston can be forced down by a screwed spindle which is turned by a handwheel. It should be noticed that the end of the screwed spindle does not bear directly on the piston, but on a thrust washer, which locates on a thrust race. This assembly is fitted in a housing which is screwed on to the end of the actuator spindle, and locked by a grub screw.

The pictorial view gives some indication of how the piston is secured to the spindle. A washer 25 dia. by 3 thick is fitted over the spindle end, and bears on a shoulder on the spindle. The top spring locator, with a recess for an 'O' ring, is next put into position, and finally the piston and diaphragm which rests on the spring locator, is secured by a 42 dia. nut, which is tightened on to the diaphragm. The thrust race housing referred to above bears on this nut.

The bottom of the cylindrical portion of the housing has a central hole tapped M27 x 3, to accommodate a screwed sleeve, the upper end of which bears on the bottom spring locator. Tension in the spring is adjusted by this sleeve, which has spanner flats machined on its bottom end for this purpose.

For the completed drawing refer to page 138

VALVE ACTUATOR

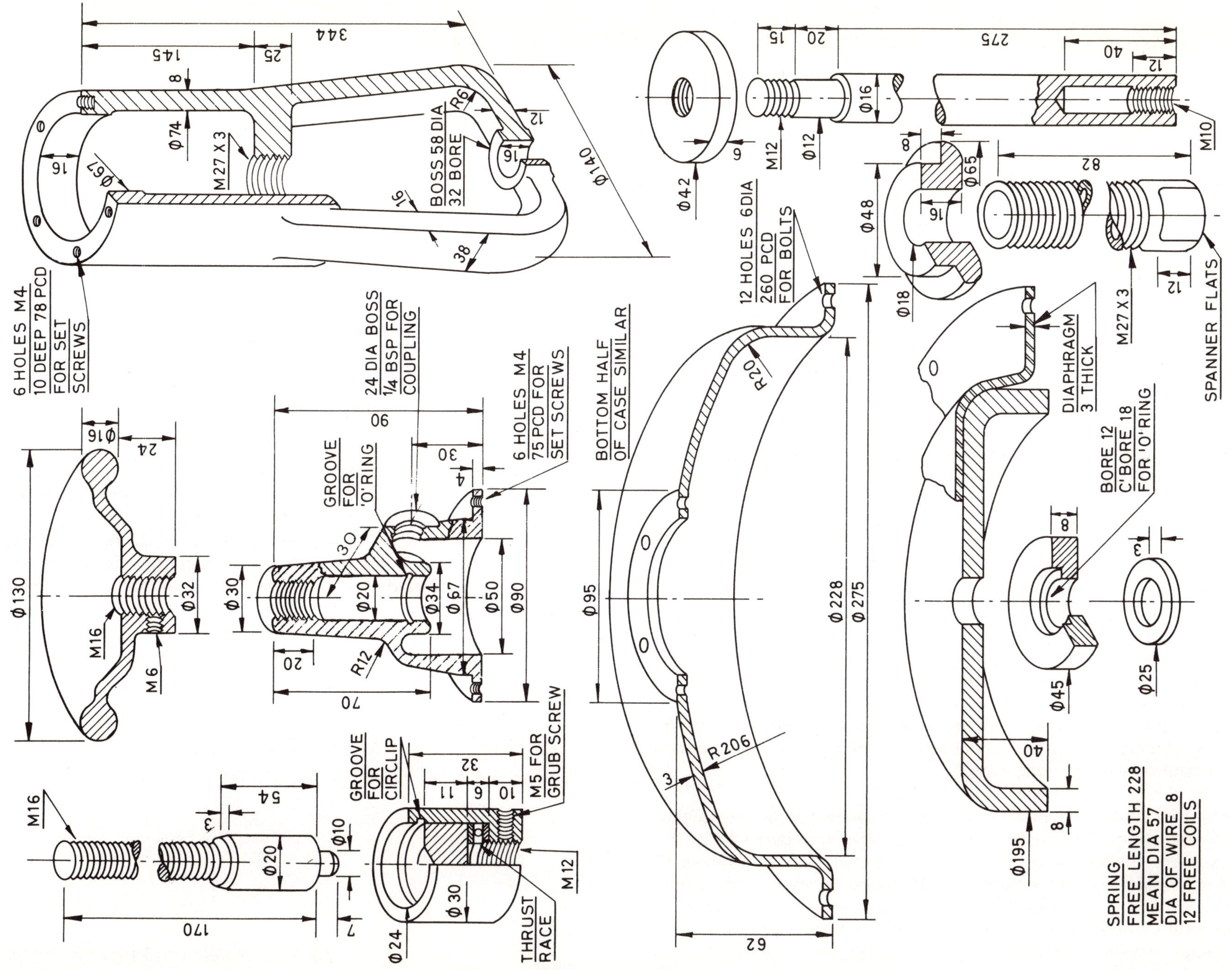

Generator Pedestal Bearing

DRAW

(i) Front elevation in section through longitudinal centre line showing bearing assembled.
(ii) End view, one half to be a section through top brass stop, the other half an outside view.

INCLUDE ON THE DRAWING A LIST OF MATERIALS.

The bearings for large generators are not carried on the end plates but on pedestals, which together with the field magnet frame, are bolted to a common bedplate. Such a pedestal bearing is illustrated pictorially on the opposite page.

The casting is hollow, with a horizontal web 12 thick, which forms the bottom of the oil reservoir below the bearing. The bottom half of the bush rests on three fitting strips, and the upper part of the casting is extended beyond the fitting strips to form pockets. The bush is entirely within the casting, and oil from the ends of the bearing runs through three 6 dia. holes in each end, into the pockets, returning to the oil reservoir through slots 38 wide. Two oil rings resting on the shaft, hang in the spaces between the fitting strips, and 19 wide slots are cast in the top brass to accommodate these rings. Lightening holes 180 x 100 in the lower part of the casting, give access to the oil drain plug, which is screwed into the horizontal web. The cap, which also has three fitting strips for the top brass, is secured to the pedestal by four M20 studs, and relative movement between the cap and the pedestal is prevented by two 12 dia. fitted dowels. A 14 dia. hole in the centre of the cap, takes a 12 dia. stop, screwed into the top brass, and this prevents rotation with the shaft. An oil hole 140 x 64 is cast in the cap, and this is fitted with a hinged cover, details of which are shown in the pictorial view.

Around a vertical centre line, draw in the sectional elevation of the pedestal required for part (i) of the question. The pictorial view shows half of this required section, and as the other half is similar, no difficulty should be experienced. The top face of the pedestal 380 above the base, is the centre line of the bush, and both halves of this could now be shown in section. Note the whitemetal lining in both brasses, with a 6 wide groove, and three 6 dia. drain holes towards each end. One of these drain holes at each end will be on the centre line, so these will be shown in this view. In the top brass, there are two slots 19 wide for the oil rings, and the sides of these slots are whitemetal lined to a depth of 1.5 leaving 16 to accommodate the oil rings. Complete this view by drawing in the cap and oil hole cover from the details given, and the $5/8$ BSP drain plug in the bottom of the reservoir could also be shown.

As the pictorial view shows sections in two planes, the half sectional end view required in part (ii) of the question together with the half outside view, can now be attempted. A local section of the oil hole in the cap is given, so this can be shown in the sectional view. The hinge block, and hinge pins for the oil hole cover will be seen in the outside view. When drawing the whitemetal lining in both brasses in the sectional end view, note that the whitemetal is cut away in each brass, for part of its length, to a width of 20. This allows the oil easier access to all parts of the bearing.

For the completed drawing refer to page 139

GENERATOR PEDESTAL BEARING

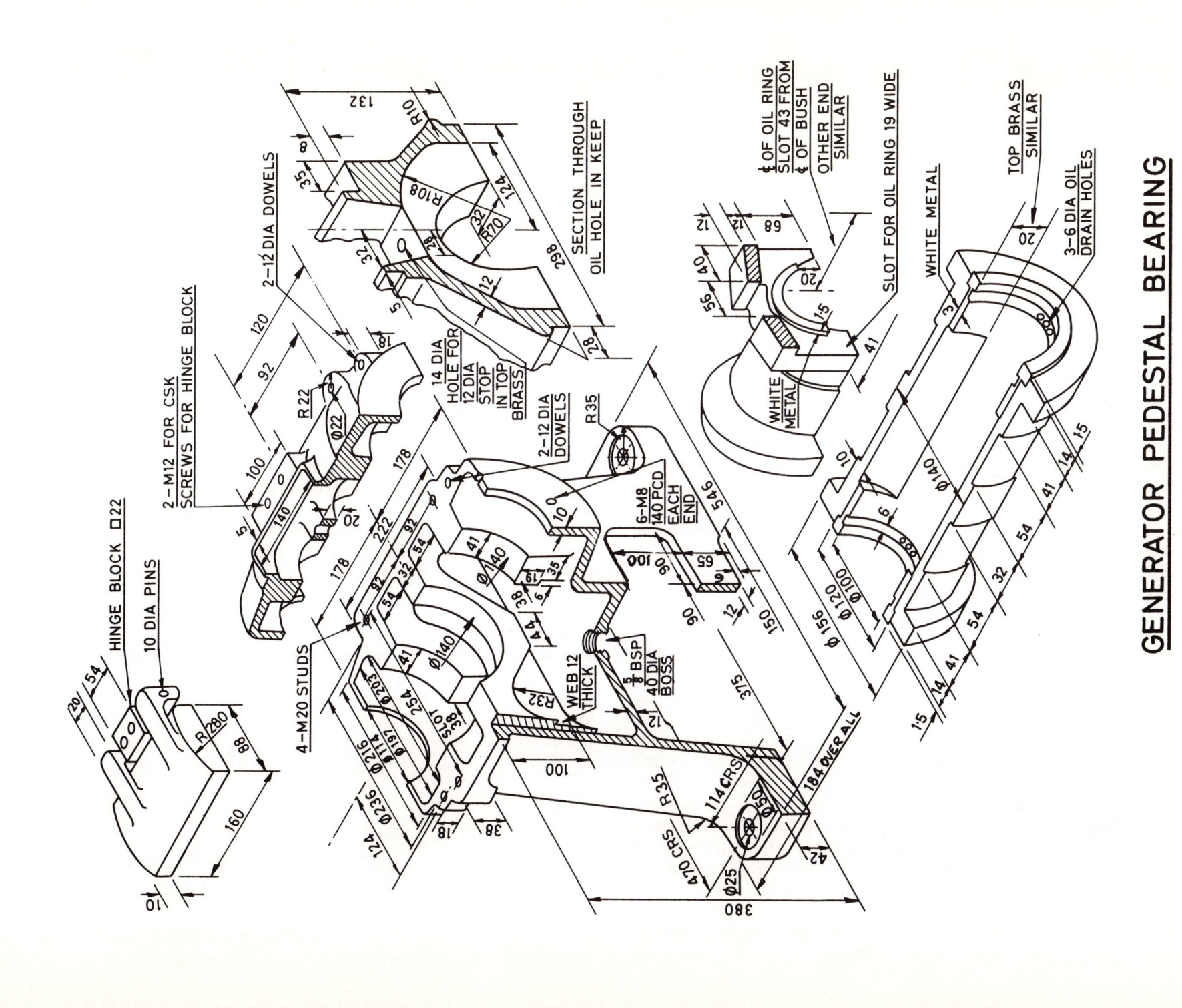

Turbine Flexible or Double Claw Coupling

DRAW

(i) Sectional elevation showing all parts assembled.
(ii) End view looking on sleeve joint.

INCLUDE ON THE DRAWING A LIST OF MATERIALS.

When the turbine rotor heats up from the cold condition, it will expand laterally, and were there no sliding coupling between the rotor and pinion, undue stresses would be put on the teeth of both pinion and main wheel, as these teeth are of the double helical type, allowing for no relative longitudinal movement. Whilst every effort is made to have rotor and pinion in perfect alignment, slight discrepancies can be taken up in the flexible coupling. With modern hobbing machines, operating in air conditioned shops to maintain a constant temperature, it is possible to achieve great accuracy in cutting the teeth on both pinion and wheel, but it is inevitable that slight discrepancies may occur in the helix angle, and the flexible coupling will accommodate these, avoiding over stressing the teeth.

The coupling is comprised of a claw fitted on the pinion shaft, and a similar claw fitted on the rotor shaft, both connected by sleeves having inwardly projecting claws and a distance piece, the torque being taken by a total of 24 fitted bolts 30 dia. The outer faces of the male claws have a part spherical surface, on to which the inner surface of the sleeve bears, with a clearance of 0.25. Two thrust rings are fitted inside the sleeves, and these limit the fore and aft travel to a total of 10.

Oil is supplied to the coupling from the forced lubrication system of the turbine, as can be seen from the pictorial view. Excess oil from the adjacent bearing is caught in a ring machined round the claw, and through holes drilled at 30° midway between each claw tooth, the oil passes through the coupling to the space between the teeth. From here it flows in a groove to the ahead and astern face of each tooth, and across the tooth faces by means of two grooves 4 wide x 1.5 deep, at right angles to each other.

Commence by drawing horizontal centre line, and the cones representing the end of rotor and pinion shafts, in their correct relative position of 219 apart. Build up the coupling from this by drawing on claws, nuts and keys. Note that there is a difference of 3 in the length of the taper on the spindle, and on the claw—this is for drawing-up purposes. The distance piece, the centre of which is midway between ends of shafts, can now be drawn, and then the sleeves which are bolted to this by the fitted bolts. Two of these bolts could be shown in this view. Continue this view by drawing in the thrust rings with their locking screws, and the nut-locking screws. Although the teeth on the claw are in line, it is permissible to move, say a bottom tooth through half a pitch, in order to show the sleeve teeth full in the bottom half of the drawing. Finish this view by drawing in the oil holes and grooves in the claw teeth—note these grooves are in the claw teeth only.

In the end view, the profile of the teeth in the claw and sleeve are shown. As there are 12 teeth, a 30°-60° set square is used to mark the centre lines of the claw teeth, and these are now drawn in 54 wide. The sides of the teeth are drawn parallel to the centre lines of each tooth. It can be seen from the pictorial view that there is a clearance of 2.5 between root of claw teeth and the tip of sleeve teeth.

For the completed drawing refer to page 140

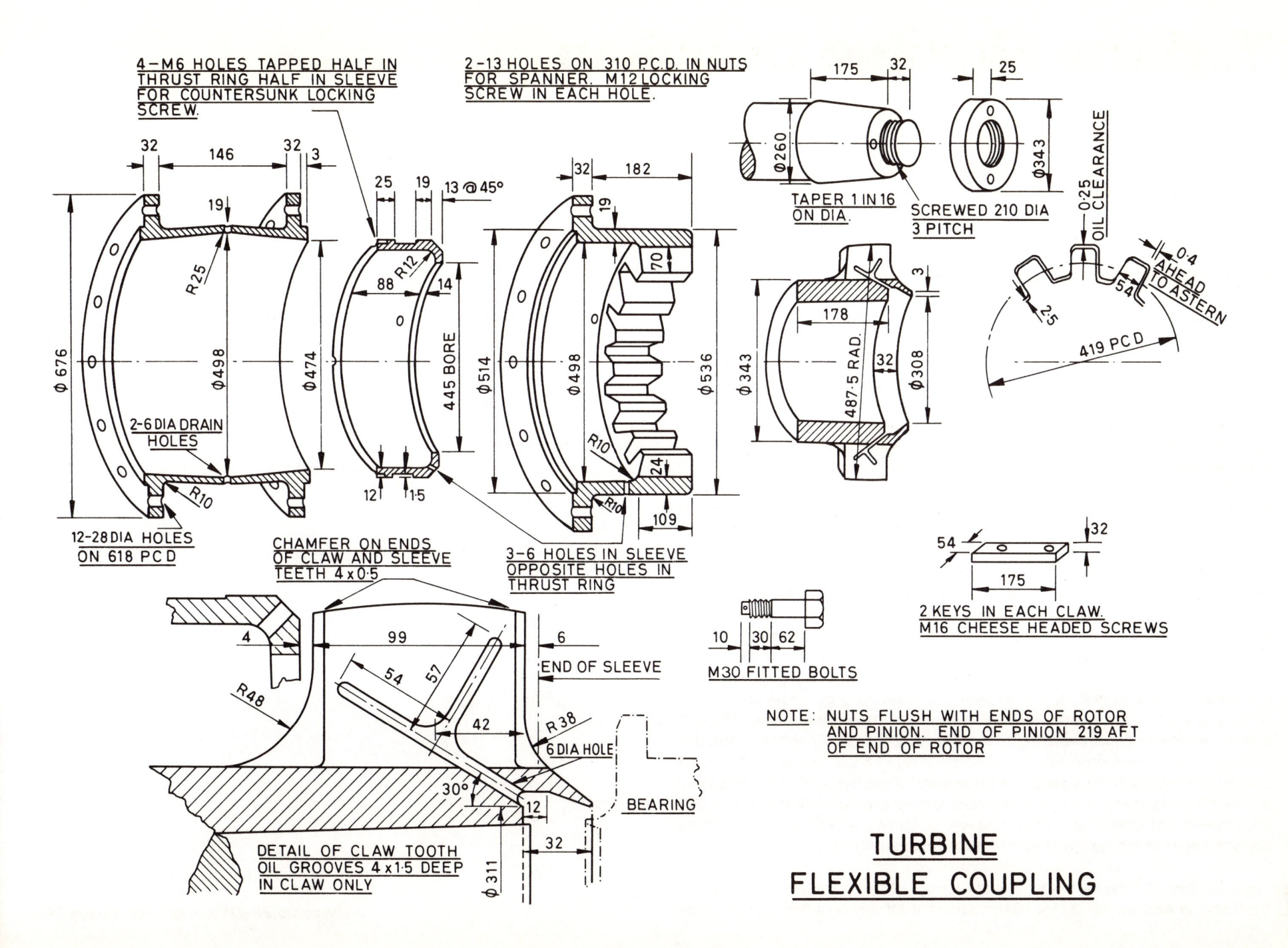

TURBINE FLEXIBLE COUPLING

Turbine Flexible Coupling

DRAW

(i) Front elevation of the assembled coupling in a longitudinal direction, the top half to be in section through the axis.
(ii) End elevation looking on the face of the flange of the male muff, showing the approximate profile of a few teeth.

INCLUDE ON THE DRAWING A LIST OF MATERIALS.

The flexible coupling illustrated opposite is a more modern version of the type on pages 80-81. As will be seen, the teeth are of involute form and of fine pitch. Both the turbine rotor and the pinion have flanges of 275 diameter formed on their ends, to which the male claws are bolted by 12 fitted bolts 25 dia. As both these claws are identical, only one is shown. The sleeves or muffs fit over these claws, and have internal teeth engaging with teeth cut on the claws. Note that there is a male and a female sleeve, so that a spigot joint is obtained where both sleeves join, and the torque is taken here by 16 fitted bolts 18 dia. Fore and aft movement is limited by a circlip, located in a groove near the ends of the teeth in each sleeve. Lubrication of the teeth is effected by 2 dia. holes drilled between each tooth in the claw.

As no dimension is given for the distance between the flange on the rotor and the flange on the pinion, the coupling will have to be built up from either the rotor or pinion end. Draw the horizontal centre line and the flange coupling of the rotor, say, as far to the left of the paper as possible, the top half in section. Next locate the flange in one claw in the recess of 256 dia. and show one 25 dia. fitted bolt. If the items are taken in the order shown in the pictorial view, the female sleeve is next drawn in over the claw, and to locate this in the correct fore and aft direction, the circlip could be shown bearing on the end of the claw teeth. Continue with the male sleeve, which fits into the 488 dia. recess in the centre flange coupling, and show one 18 dia. fitted bolt. To locate the pinion claw inside its sleeve, the inner ends of the teeth of claw and sleeve will be flush, as the position of the circlip at this end will allow the required axial movement. One 25 dia. fitted bolt will be shown securing the claw to the pinion flange, which is similar to the flange on the rotor.

The end view, looking on the face of the flange of the male sleeve, will show the flange of 488 dia. drilled for 16 fitted bolts of 18 dia. on a 446 PCD, and the bore of 375 dia. will be seen. The teeth on the muff, and the internal teeth on the sleeve will be seen in this view, and for their approximate profile reference should be made to page 18.

For the completed drawing refer to page 141

GEAR DATA

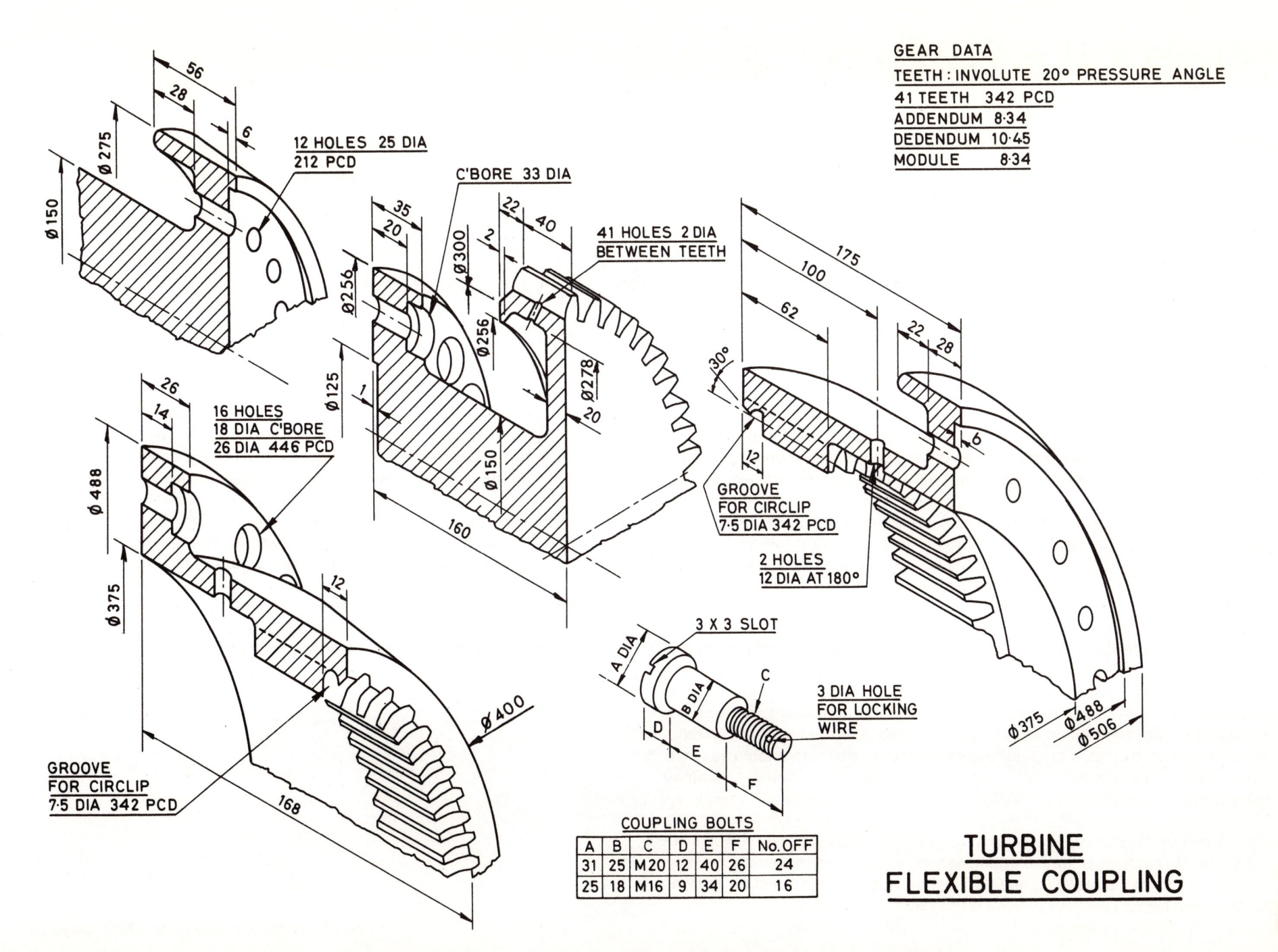

TEETH: INVOLUTE 20° PRESSURE ANGLE
41 TEETH 342 PCD
ADDENDUM 8·34
DEDENDUM 10·45
MODULE   8·34

TURBINE FLEXIBLE COUPLING

COUPLING BOLTS

A	B	C	D	E	F	No. OFF
31	25	M20	12	40	26	24
25	18	M16	9	34	20	16

83

High Lift Safety Valve

DRAW

(i) Front elevation with waste steam branch behind. One half of this elevation to be a section through the vertical centre line, the other half an outside view.

(ii) End view.

ON THE SECTIONAL ELEVATION INDICATE THE LIFT OF THE VALVE, AND ALSO INCLUDE A LIST OF MATERIALS.

A High Lift safety valve of the type illustrated pictorially on the opposite page is a common question in the drawing examination for a Second Class Certificate.

Referring to the completed drawing on page 142 the valve operates as follows. The valve lid is subjected to boiler pressure, which tends to open it against the compression of a heavy spring, the tension on which can be adjusted by a nut, screwed into a bush in the top cover. When boiler pressure exceeds spring pressure, the valve opens and waste steam escapes to atmosphere. In addition, this waste pressure acts on the under side of a piston, fitted on the valve spindle and working in a floating ring, which is held against a slotted diaphragm, also by the waste steam pressure. This additional load acting against the spring pressure, will allow the valve to lift further. The radii on both seat and lid create a nozzle effect, when the valve opens.

When the spring tension has been adjusted to the correct blow off pressure, a split ring is fitted between a collar on the compression nut, and the top of the screwed bush in top cover. As the depth of this ring will determine the amount of compression which can be put on the spring, it is important that it is not readily removed, and for that reason a cap is fitted over the top of the spindle. A cotter through the cap and spindle takes a padlock, and care should be taken to ensure that there is sufficient clearance in cotter slot in spindle, to allow valve to lift.

The easing gear consists of a shaft in two bearings cast on the cover. Forks are forged on the shaft, to engage with a collar on top cap, and on turning the shaft by a lever, the cap is forced up against the spring pressure, reducing the load on the top of the valve lid, and allowing the valve to open fully.

Commence the front elevation by drawing two vertical centre lines at 203 centres. Make say the left hand side the sectional view, and draw in valve chest from dimensions given. The seat, secured by six M12 set screws could now be located, and valve lid also drawn in. Continue the sectional part of the elevation by drawing in spindle, piston, floating ring and diaphragm, from the dimensions given. Next could be drawn the spring cases, the left hand case in section the other drawn as an outside view. The top cover with its bush secured by two M6 grub screws is now added, together with the compression nut, compression ring and top cap. With the compression nut in position, the top spring cap can be located, and then the spring. Finish off by drawing in the cotter, and the slot in the spindle to accommodate this, easing gear bearings and shaft, with an operating lever keyed to one end of the shaft.

The end view will show more detail of the waste steam branch, the dimensions for which are given on the pictorial view. An important item in this branch is the ¾ BSP drain hole, sometimes overlooked by students

For the completed drawing refer to page 142

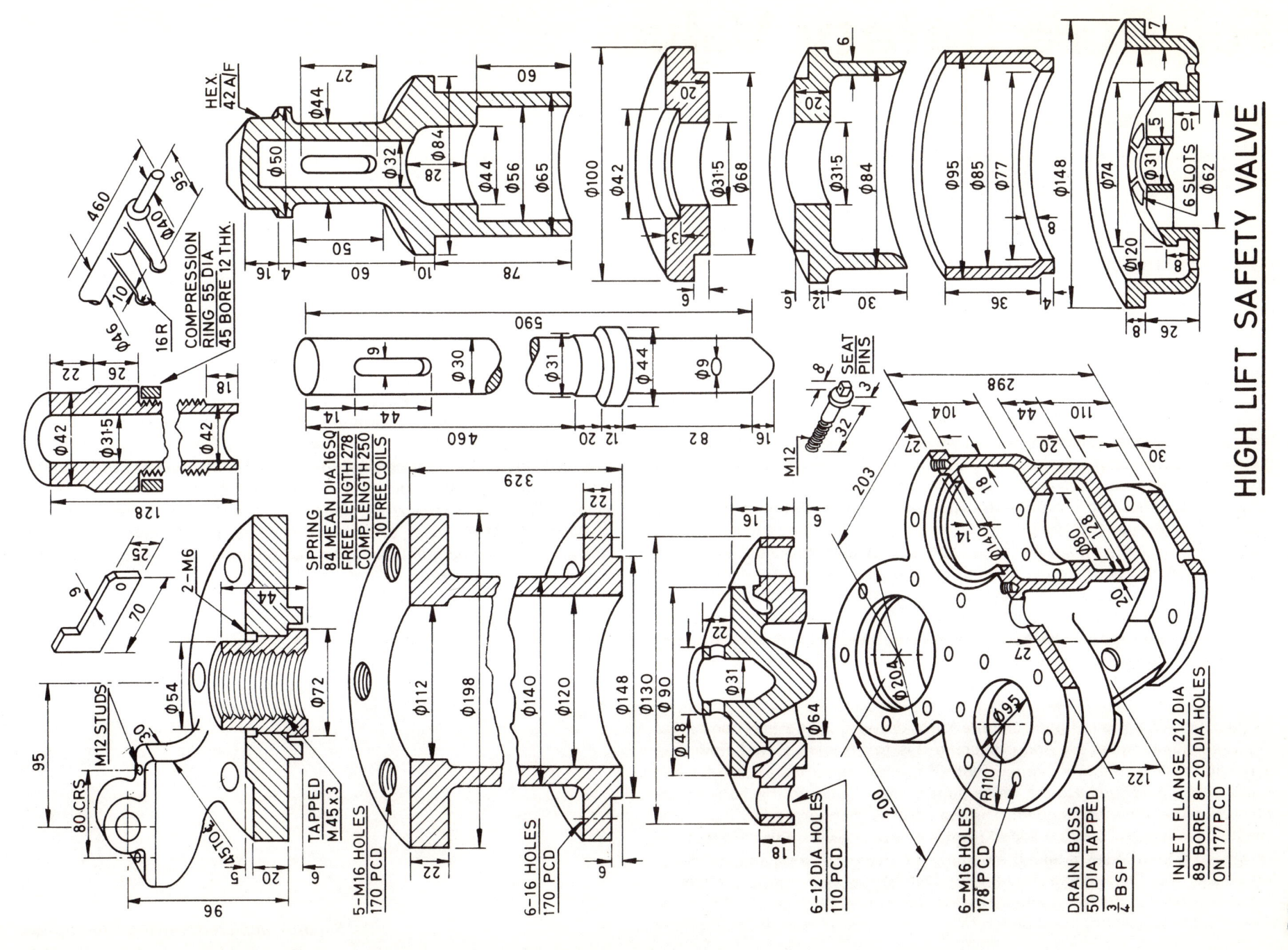

HIGH LIFT SAFETY VALVE

85

Full Bore Safety Valve

DRAW

(i) Elevation in section showing valve assembled.
(ii) Plan view.

INCLUDE ON THE DRAWING A LIST OF MATERIALS.

For boiler pressures in excess of 40 bar, the full bore valve shown opposite is fairly common. To understand how the valve functions, reference should be made to the assembled drawing on page 143. Steam at boiler pressure enters the 76 dia. left hand branch, and acts on the bottom of the valve, tending to hold it shut. Another branch 19 dia. is cast on the top cover, and this takes steam from the outlet of a pilot valve, also mounted on the boiler. When the blow off pressure is reached, this pilot valve lifts, and allows steam to enter the space above the 108 dia. piston. As this piston is larger in dia. than the valve, the load forcing valve down, will be greater than the load tending to keep it shut, hence the valve opens, and allows waste steam to escape to atmosphere, through the 152 dia. right hand branch. The amount of valve lift, is controlled by the length of the conical projection cast on the bottom cover. The comparatively light spring, fitted above the top cover, is merely to assist the valve closing with a snap action. The easing gear consists of a shaft, fitted in two bearings cast on the top cover, with a trigger forged on the shaft, and engaging with the end of valve spindle. When the shaft is turned in the proper direction, this trigger forces the valve spindle down, and hence lifts the valve off its seat.

To draw the two views required, commence by drawing the vertical centre line of the sectional elevation, and producing this to the plan view below. The horizontal centre line in elevation, will of course be through the inlet and waste steam branches. Round these centre lines, build up the body of the valve from the dimensions given. The top cover is then located by its spigot, and one easing gear bearing added from the details given. Next draw in the valve seat, the cylinder to take the piston, and then the valve and spindle. It will be seen that the piston is made in two, the bottom half being a force fit on the spindle, whilst the top piston screws down on this. A locking pin could be shown in the top piston, and a piston ring on each piston, as indicated on the pictorial view. The neck bush is added to the top cover in way of the spindle, and top nut, spring, and spring housing now drawn in. Finish off the easing gear by adding the trigger, easing gear shaft, and two M20 nuts on the bearing keep. The bottom cover completes this view. The valve and spindle should not be shown in section, except for a portion of the valve locally, in order to show the stellite deposit. There is a similar deposit on the valve seat.

The plan view is fairly straight forward, as no hidden detail need be included. The widths in plan view are projections from the elevation, and the various diameters seen in this view can also be obtained from the elevation. Five M20 studs secure the top cover, and when these have been drawn in plan view, the one seen in the sectional elevation can be obtained by projecting from the plan view to the elevation.

For the completed drawing refer to page 143

FULL BORE SAFETY VALVE

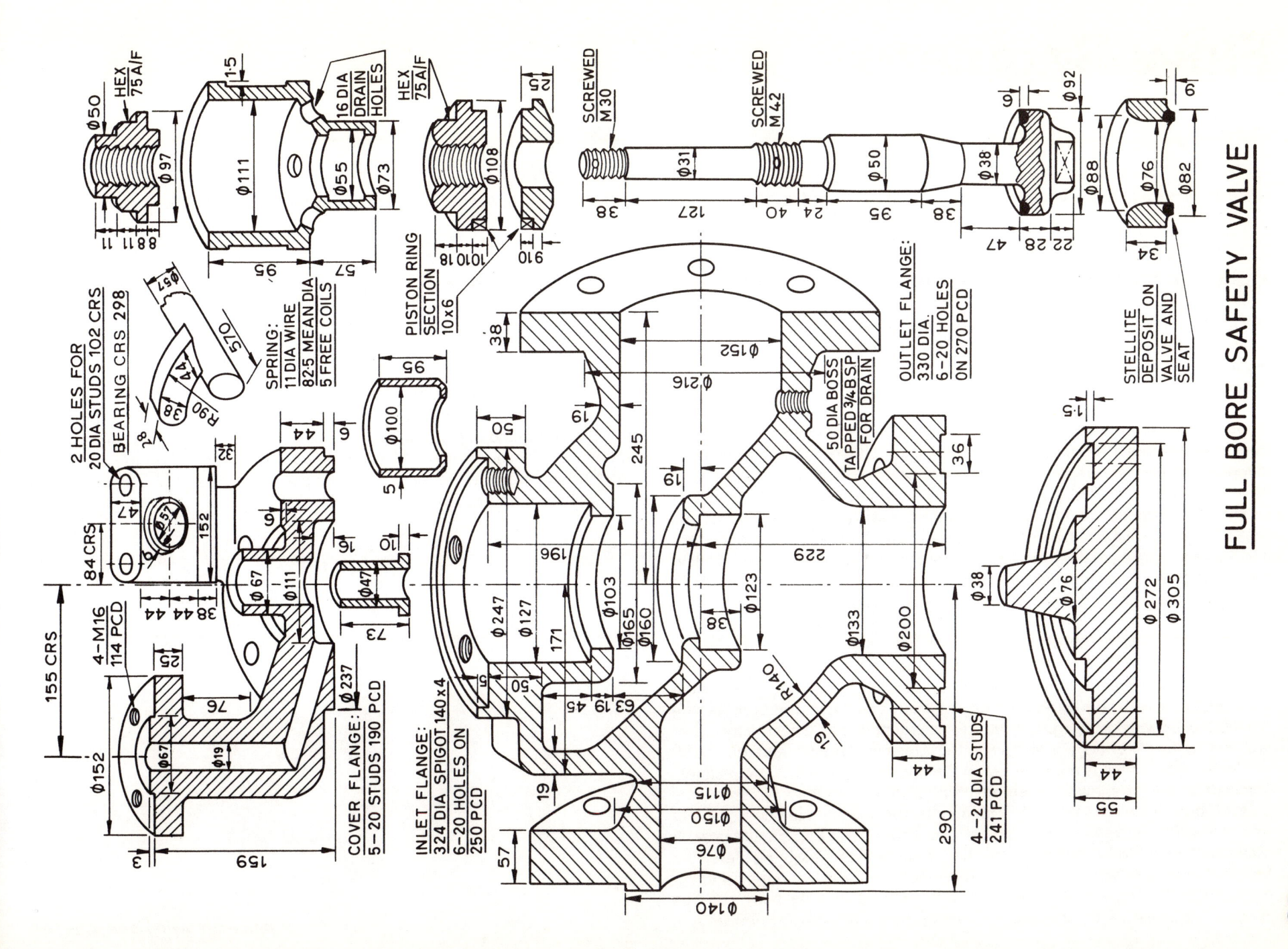

Flow Regulator

DRAW

(i) Front elevation in section through inlet and outlet branches, showing regulator assembled.
(ii) Plan view, one half an outside view with top cover and valves removed, the other half a section through centre line of branches, with valves removed.

INCLUDE ON THE DRAWING A LIST OF MATERIALS SUITABLE FOR A REGULATOR USED FOR SEA WATER.

The flow regulator shown pictorially opposite is automatic in operation, and is designed to give constant flow conditions on the downstream side of the regulator. One marine application of such a regulator would be in the feed line of an evaporator.

A variable orifice is fitted in the inlet branch, and a ¼ BSP tapping is taken from a point before this orifice, and led to the space above a pressure sensitive diaphragm. If the inlet pressure be P_1, and the pressure after the orifice be P_2, the nett pressure acting down on the diaphragm will be $P_1 - P_2$, and this pressure determines the position of the valves relative to their ports. Should the outlet pressure decrease (indicating an increased flow), $P_1 - P_2$, increases, and the valves are moved towards the closed position, so that the original flow rate is restored. The downward forces acting on the diaphragm are balanced by a spring, fitted under the valve stem in the bottom cover.

Even though students may not be familiar with this type of regulator, its assembly should not prove difficult, as the various items are, to a large extent, shown in their correct relative positions in the pictorial views. Draw in the horizontal and vertical centre lines for the front elevation, and build up the regulator body with its top and bottom covers around these. Only the top cover is shown in the pictorial view, but the bottom cover is identical, and the ¼ BSP port in the side, which is not required in the bottom cover, would be plugged. The valves could be shown in their position for maximum flow, which means that the stop, driven into the ¼ BSP plug in the top cover, would be in contact with the end of the valve stem. The stem, and both valve plugs could now be drawn in, together with the piston, the diaphragm and the retaining plate. The flange of the diaphragm is sandwiched between the top cover flange and the body, and an 'O' ring, fitted in the groove in the top flange of the body, could be shown. The valve ports, screwed externally 2 BSP are next drawn in, and their locking rings indicated. The spring holder, spring, and butterfly valve fitted in the orifice ring, completes this view.

A local sectional plan view of the inlet branch is given, so that its shape can be indicated in the plan view required for part (ii) of the question. The butterfly, spindle, and gland are also shown pictorially, and these items can now be drawn in the sectional plan. A handwheel is normally fitted on the end of the butterfly spindle, together with an index plate, but these have been omitted from a point of view of time.

The completed drawing is on page 144

FLOW REGULATOR

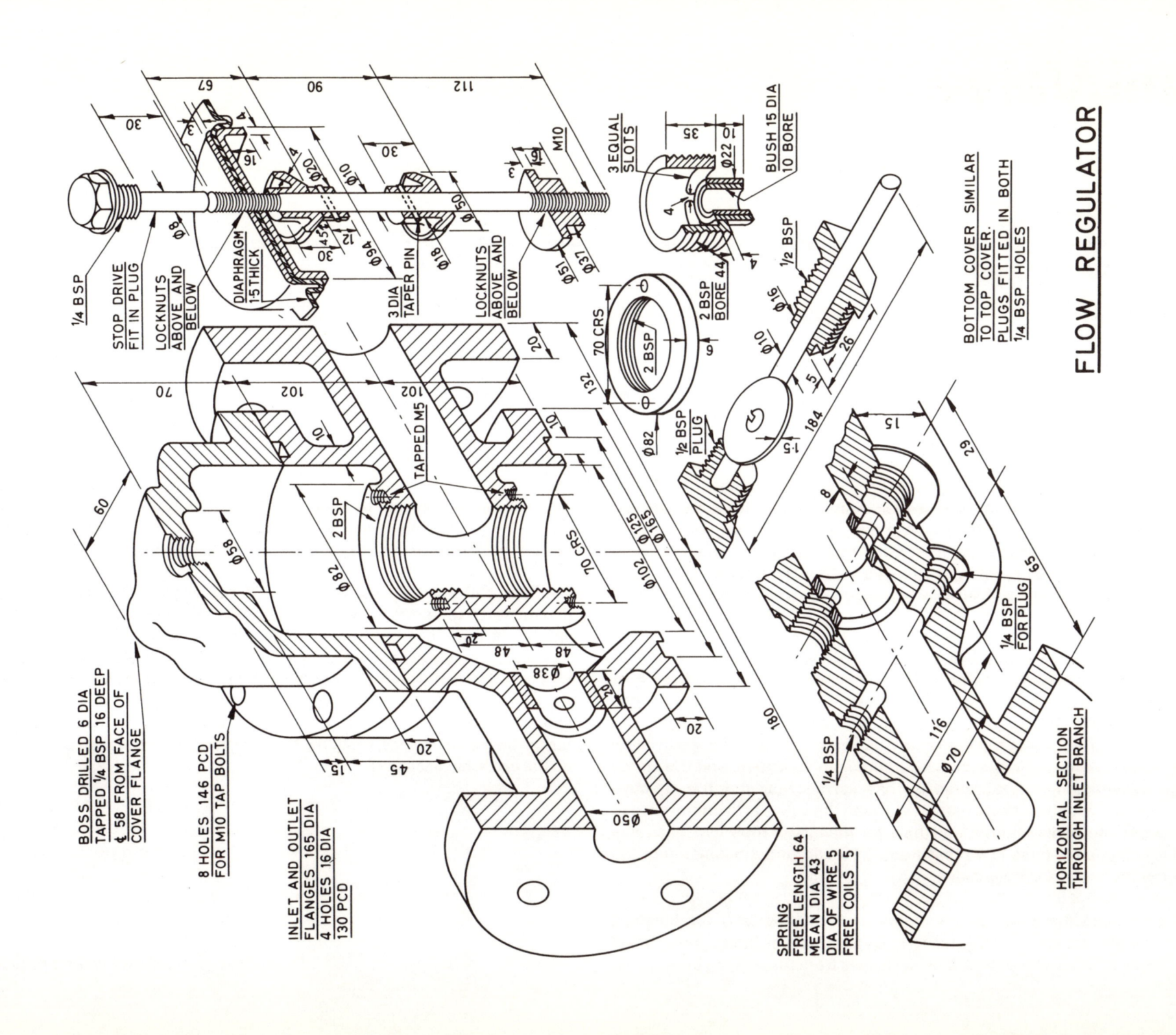

Fuel Valve

DRAWING No. 1

DRAW

(i) Valve body in section, with fuel inlet connection at the back, the plane of the section to be taken through the vertical centre line.
(ii) Plan view of valve body projected from (i).

DRAWING No. 2

DRAW

Sectional elevation of valve, showing all parts assembled. The view should be taken through the vertical centre line, with fuel inlet connection at the back. Indicate the lift of the valve.

INCLUDE ON THE DRAWING A LIST OF MATERIALS

Should this question appear in the Drawing examination, the views required would probably be the assembled view asked for in Drawing No. 2, with principal dimensions inserted, together with a plan view. For students who are not familiar with the assembly of a fuel valve, however, dimensions have been omitted from the assembled view on page 145, from a point of view of clarity.

Fuel at high pressure enters the body of the valve through the 1 BSP connection, shown clearly in the pictorial view opposite. Through a 6 dia. port drilled at 4½° to the vertical, the fuel reaches the valve spindle sleeve, and through a 4 dia. port drilled in this component, enters the nozzle holder. The nozzle, which is shrunk into the holder, is relieved on its upper flange to allow the high pressure fuel to enter the bore of the spindle sleeve, and thus force open the valve against the compression of a spring. Note that the effective area on which the high pressure fuel acts to open the valve, is the difference in area between the 14 dia. spindle, and the 6 dia. flat face of the valve. The valve shown is designed to open at a pressure of 215 bar, and atomisation of the fuel is effected by three 0.9 dia. holes.

A port, similar to the inlet port, is drilled in the valve body for bleeding purposes. At the upper end of this port a ball valve 8 dia. is fitted, and to bleed the fuel valve this valve is opened, and fuel issues from the leak-off port.

The valve lift is adjusted by a sleeve screwed ⅜ BSP, which passes through the spring compression plug having an external thread 1½ BSP, whilst a 6 dia. rod screwed into the spring holder acts as a lift indicator.

The faces of the spindle sleeve and nozzle holder are lapped to a mirror finish, and the valve spindle is lapped into the sleeve, so, consequently, if a new valve spindle is required, the sleeve must also be renewed. Despite the fine tolerances, however, some leakage will occur, and to deal with this a 6 dia. port is drilled in the body. This port, at 25° to the ₵, is drilled into the bevel at the bottom of a ⅜ BSP hole which is parallel to the ₵. A whitemetal ring, sandwiched between a bevel on the end of the nozzle cap and a similar bevel turned on the valve body, prevents fuel which may have leaked past the thread on the cap, from escaping.

For the completed drawing refer to page 145

FUEL VALVE

Tunnel Bearing

DRAW

(i) Sectional elevation through axis of shaft, showing bearing assembled on shaft.
(ii) End elevation looking on aft end of bearing.
(iii) Half section looking aft, the section plane for the bottom half to be taken through centre of pads, and the section plane for the top half to be taken through the centre line of the top casing.

INCLUDE ON THE DRAWING A LIST OF MATERIALS.

The bearing illustrated on the opposite page is a fairly common type in both naval and merchant ships, and with the exception of the aftermost bearing, which has pads on both top and bottom halves, it is used on the intermediate shafting between the thrust block and the stern tube. The number required of course, would depend on the length of shafting.

The bottom half of the shaft is supported by three pads, making a total included angle of 168°, so that the upper face of the top pad, on either side, is 6° below the horizontal joint. Four cheese headed stop screws screwed into webs in the top cover, project down into this space and are fitted so as to allow the pads slight movement to take up their own position on the shaft. In the bearing illustrated, this clearance is given as 4 for the three pads. Lubrication is effected by a thrower ring clamped to the shaft, and running in an oil sump cast in the bottom half. An oil scraper rests on the top of the thrower ring, and deflects the oil picked up on to the top of the shaft, and hence into a groove 28 wide x 20 deep behind the pads. Slots in the pads allow the oil access to the bearing surface. Any oil which travels along the shaft towards the ends of the bearing, is prevented from escaping by a deflector, not shown, fitted into a boss on each end. Doors are provided at each end for cleaning purposes, and in addition a ½ BSP drain plug is fitted. A cooling coil is often fitted in the bearing sump, but this has been omitted.

Commence by drawing a common horizontal centre line for the three views, and arrange the views with the sectional elevation in the centre, and an end view on each side of it. The half section of the bottom half looking aft is quite clear from the pictorial view, as the section given here is taken through the centre of the pads. A local section of the top half in way of one centre joint bolt is also shown, and this will be the required section for the top half.

Both top and bottom halves of the bearing are sufficiently illustrated in the pictorial view, for the sectional elevation to be drawn in, with the thrower ring clamped on the shaft. Two of the pad stop screws can also be shown in the elevation, and one in the sectional end view, with a clearance of 4 between stop screw and pad flange. The end view could be brought along with the other views, as time can be saved by using the same compass setting in all views, before it is altered.

For the completed drawing refer to page 146

TUNNEL BEARING

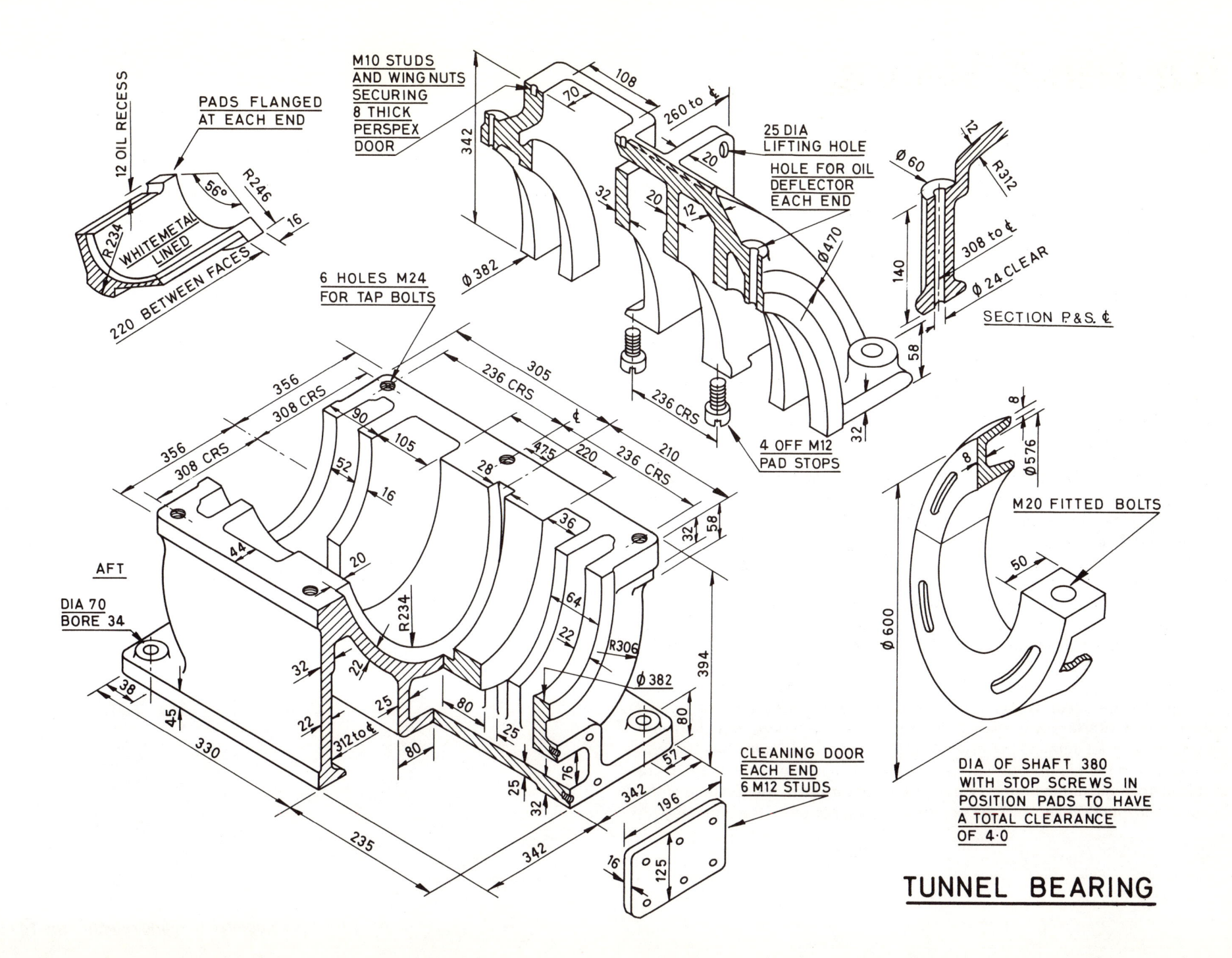

Air Inlet Valve

DRAW

(i) Front elevation of the assembled valve in section with inlet branch on the left. The section plane should be taken through the centre of the valve and through the right hand holding down stud hole.
(ii) End elevation.
(iii) Plan view.

INCLUDE ON THE DRAWING A LIST OF MATERIALS.

The air inlet valve shown pictorially on the opposite page is from a four cycle pressure charged direct drive main engine, a type of engine no longer being built. It should be remembered however, that the questions set in the DoT Engineering Drawing examination do not necessarily conform to the most up to date marine engineering practice, and from a drawing point of view this example is worth attempting.

The valve and valve body should not be difficult to assemble; the 60 dia. valve stem fits in the 65 dia. central hole in the valve body, and the mitre on the valve head locates in a similar mitre on the seat which is shown enlarged in a local section. The valve guide takes the form of a piston fitted with three rings, which works in a cylinder 110 dia. cast on the top of the body. A recess is turned in the valve stem near the end, and a split collar 75 dia. x 19 deep is fitted in this recess, and bears on a shoulder inside the piston. The outside dia. of the piston is increased from 110 dia. to 130 dia., and the shoulder thus formed bears on a split spherical ring. This ring in turn is located in the spring keep plate, which bears on the ends of both springs. Two springs are used, one wound to a right hand helix, and the other to a left hand helix. Opposite hand springs are used, so that in the event of a spring breaking, the broken spring cannot interfere with the intact spring. The piston rings which prevent air leakage are not detailed in the pictorial view, but these could be shown in the grooves in the piston. For lubrication of the piston eight holes 2 dia. are drilled at an angle at the bottom of a 5 x 5 groove turned in the top face of the housing and four drain holes 4 dia. are drilled at the bottom of the cylinder.

For the completed drawing refer to page 147

AIR INLET VALVE

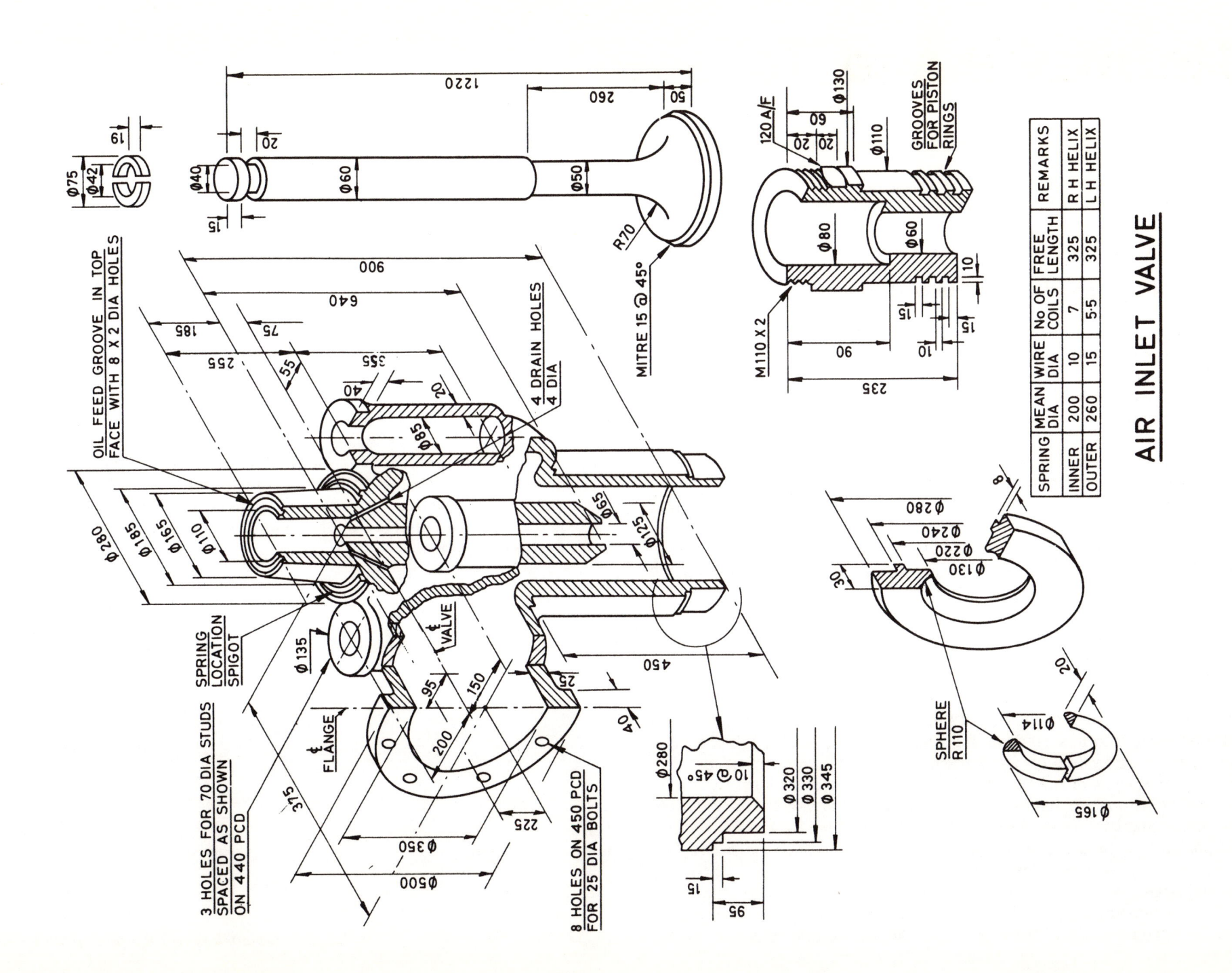

SPRING	MEAN DIA	WIRE DIA	No OF COILS	FREE LENGTH	REMARKS
INNER	200	10	7	325	R H HELIX
OUTER	260	15	5.5	325	L H HELIX

Sterntube and Tailshaft

DRAW

(i) Sectional elevation through axis of shaft, showing all parts assembled.
(ii) Detailed views of oil gland to a larger scale.

INCLUDE ON THE DRAWING A LIST OF MATERIALS.

For many years it was standard practice to fit water lubricated, lignum vitae-lined bushes in the stern tubes of ocean-going ships, and although there are many ships still afloat with this arrangement, new tonnage generally has some form of oil lubricated sterntube. The arrangement shown pictorially on the opposite page illustrates a typical oil-filled sterntube.

It will be seen that there are two bushes in the tube, a long bush at the aft end, and a shorter neck bush towards the forward end. These bushes are generally made of cast iron, and lined with whitemetal which is keyed both longitudinally and circumferentially. The gland at the forward end is similarly lined with whitemetal, and all the liners have oil grooves cut parallel to the axis of the tailshaft. Lubrication is generally by gravity from an oil tank in the engine room, and situated above the load line so as to ensure oil entering the sterntube under all conditions of draught. Special oil for this duty is supplied by all the leading oil companies.

A special feature of this type of sterntube is the oil gland at the aft end, which is sandwiched between the forward face of the propeller boss and a cast iron wearing ring, secured to the flange of the aft bush by countersunk screws. It should be noted that this gland is not attached in any way to the propeller, shaft or sterntube so that it is free to revolve, remain stationary or to creep. The inner diameters of the gland rings are turned slightly larger than the diameter of the shaft, so that there is a small oil film between rings and shaft, on which the gland floats. An oil-resisting rubber member between both rings gives flexibility to the gland, and keeps the rubbing faces close together. Two drivers located in suitable slots in their opposite gland ring ensure that no torsional strain comes on the rubber member. This gland can be removed or fitted without disturbing propeller or coupling bolts. The slots shown around the outer circumference of both halves are for temporary bolts, which give compression to the rubber member when fitting. When the gland is in position on the shaft, the 30 dia. joint studs are inserted, and the compression bolts withdrawn.

The views required for this exercise should not cause any difficulty, as all the important dimensions are given, or can be found. Notice that for drawing-up purposes, the propeller boss is 24 longer than the taper on the tailshaft. The correct fore and aft location of the tailshaft in the stern tube is found by subtracting the oil gland compression of 6 from its free length of 186, and this will be the distance between forward end of propeller boss and aft face of wearing ring.

The completed drawing appears on page 148

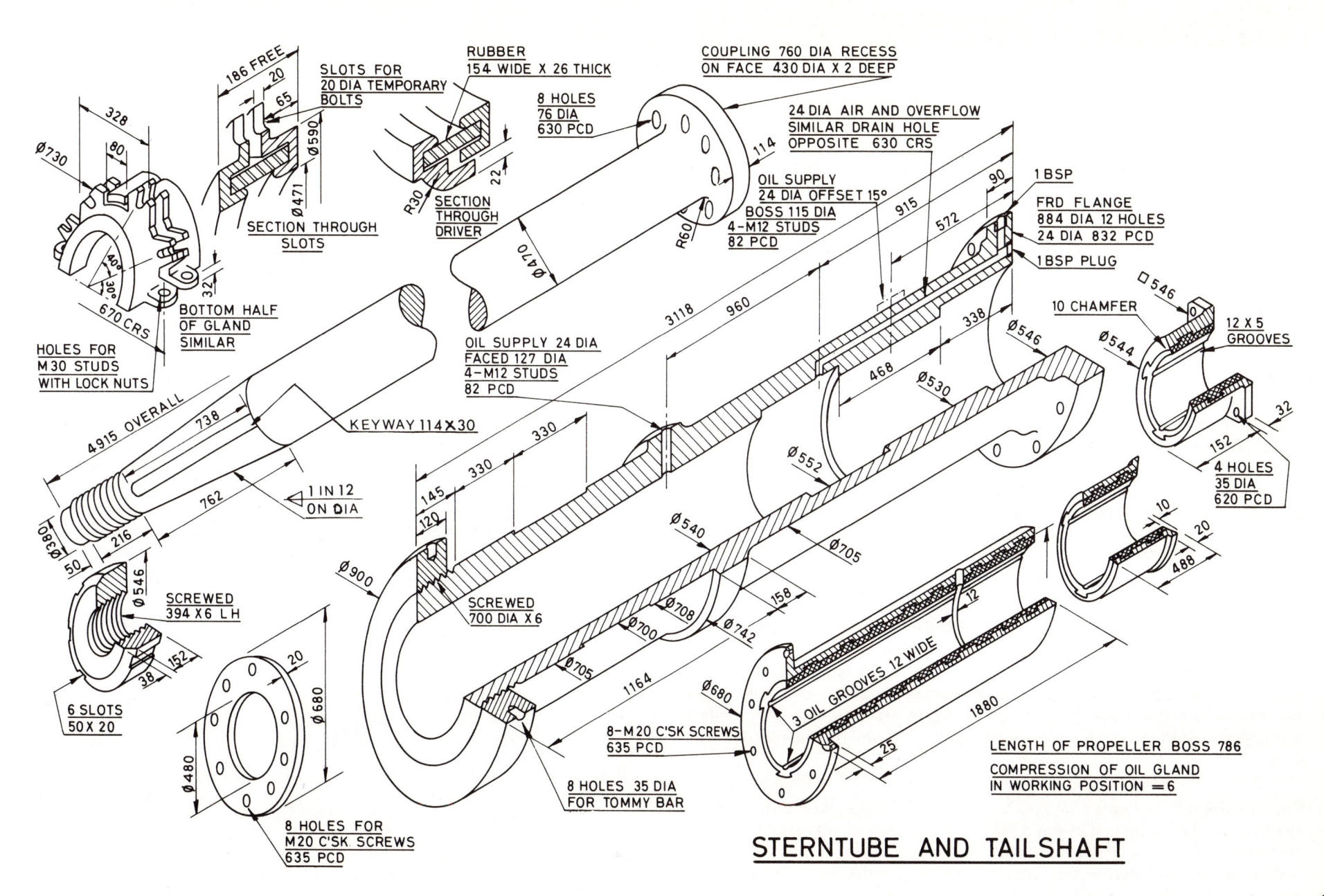

STERNTUBE AND TAILSHAFT

97

Michell Thrust Block

DRAW

(i) Front elevation in a longitudinal direction, one half in section through vertical centre line, the other half an outside view.
(ii) End view, one half in section through a journal, the other half a view looking on the face of one set of thrust pads.

NOTE: LINERS OF TOTAL THICKNESS 11 TO BE FITTED ON THE BACK OF EACH SHOE, TO GIVE A FORE AND AFT CLEARANCE OF 1.0

INCLUDE ON THE DRAWING A LIST OF MATERIALS.

Before commencing to draw, the pictorial view should be studied carefully, so that a clear picture is obtained of the assembly of the thrust block. The thrust shaft is given as 408 dia., on which is forged a collar of 854 dia. and 127 thick, with a fillet radius of 12. Against both the forward and aft faces of this collar, whitemetal-lined thrust pads are fitted to take the thrust in the ahead and astern directions. These pads are fitted in a 138 wide groove in a shoe, with a clearance of 3 radially, and as the sector angle of the shoe is given as 270°, and the included angle over the pad stops 45°, there must be six pads on each side of the collar. These would be arranged four below the horizontal centre line, and one port and starboard above the centre line. A journal bearing, whitemetal lined, is provided at each end of the thrust block.

From the Engineering Knowledge point of view, certain items are included on the completed drawing, but as a drawing exercise to be completed in six hours, these need not be shown. Amongst these items is a cooling coil, the couplings for which are fitted in the circular door in the oil sump. Oil deflectors are fitted beyond the ends of the journals, to return to the sump any oil tending to escape along the shaft. An oil level gauge is also fitted on the side of the oil sump.

The journals are lubricated by means of an oil scraper, which deflects oil picked up by the thrust collar into a reservoir cast in the pad stop. From here the oil flows through a 25 dia. hole into the top bush of the journal. A film of oil is also maintained between the faces of the pads and the thrust collar, as the pads are free to tilt slightly, allowing a wedge-shaped film of oil to form between the collar and the face of the pad.

Commence the elevation by drawing in the thrust shaft, and make, say, the right hand side the sectional view. Produce the horizontal centre line to the end view, and draw in three pads in this view—two below and one above the horizontal centre line. As in the elevation, make the right hand side the sectional view. Project the bottom pad across to the elevation, and draw this pad at 64 thick just clear of the collar. The shoe can now be located in the sectional elevation, and drawn in, with liners to a total thickness of 11 behind it. The right hand journal can now be indicated in section, and the bottom half of the block finished off in both views from the dimensions given. Next draw in the top of the casing, cover, and pad stop in the end view, and project across to the elevation for the respective heights. Finish off the elevation by showing half of the oil scraper located on the top of the thrust collar, and deflected into the reservoir cast in the pad stop.

For the completed drawing refer to page 149

MICHELL THRUST BLOCK

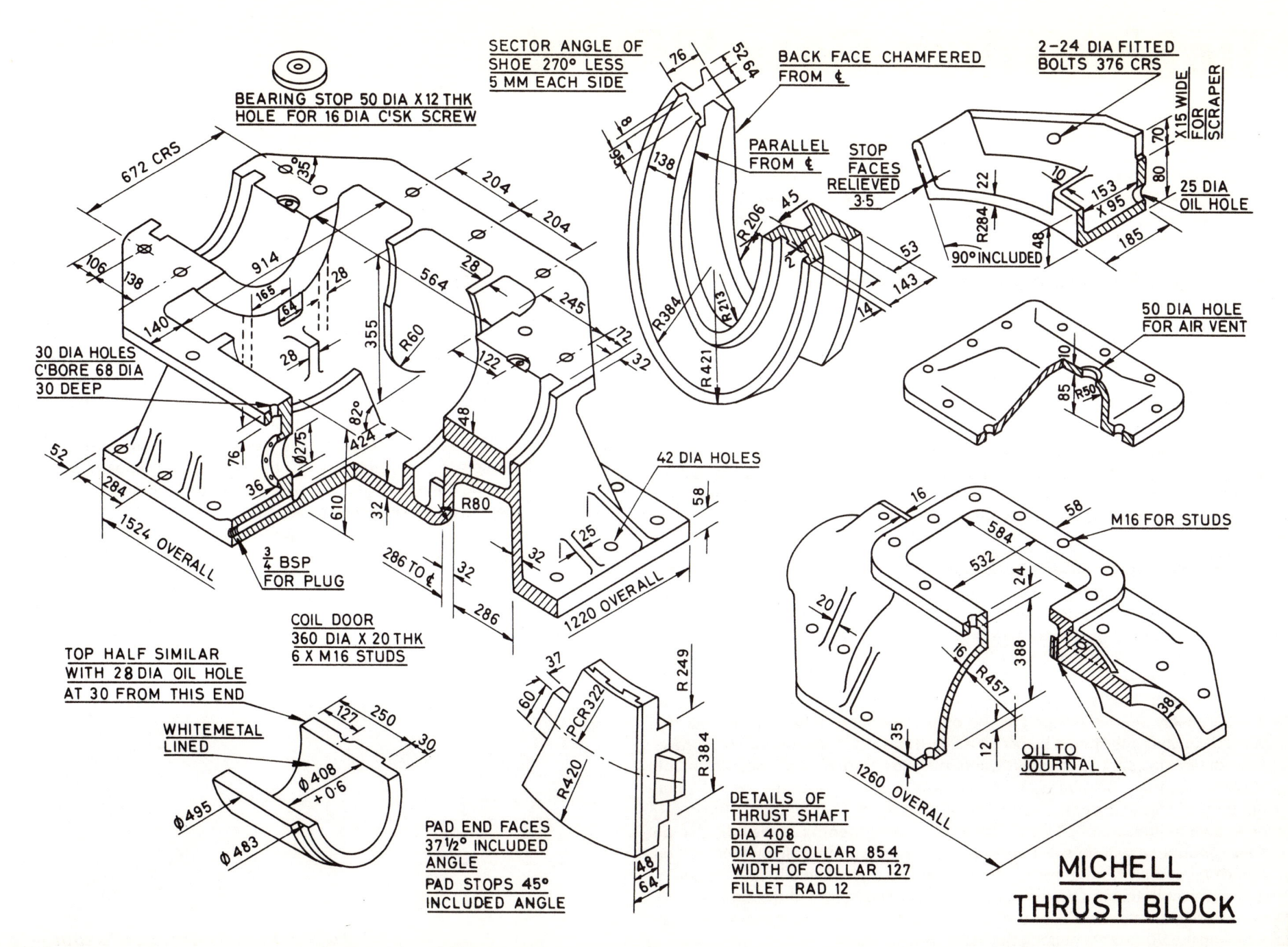

99

Turbine Main Gear Wheel

DRAW

(i) Elevation in section through axis of shaft showing wheel assembled.
(ii) End elevation one half in section through ℄ and other half an outside view.
(iii) Draw the approximate profile of a few teeth to a larger scale than the main drawing.

INCLUDE ON THE DRAWING A LIST OF MATERIALS.

Main gear wheels on modern turbines are either welded or built up, these designs having superseded the cast wheel, with shrunk on rim. The wheel illustrated pictorially on the opposite page is of the built up type, and would be used in a double reduction train.

The assembled wheel comprises a hub, a rim on which the double helical teeth are cut, two 42 thick side plates and a forged truncated cone, bolted internally by its flanges to the side plates. A thrust collar would also be forged on the aft end of the hub but this is not shown.

Having decided upon a suitable scale, draw in the horizontal centre line for the sectional elevation, and produce to the end view. Draw in the hub from the details given, and locate both side plates by the spigots on the hub. Notice that the outer periphery of the aft plate is machined to take the 78 deep flange, forged on the rim, and the inner diameter machined to take the cone flange, whilst on the forward plate, the cone flange fits into a machined recess at the outer diameter. This may not be obvious from the pictorial view, as the cone has purposely been illustrated in the opposite way to which it fits into the wheel, but students should be prepared for this kind of deviation. The cone is now drawn in, sandwiched between the side plates, and located in the recesses in the side plates, as explained above. There are six 500 dia. lightening holes in the cone, and as two are cut by the section plane, these can be shown in the section. Two of the other four lightening holes will also be seen, and these will be seen as ellipses, the construction for which can be found on page 12. Next draw in the rim in section from the dimensions given, and show teeth top and bottom, not in section. The height of the teeth can be calculated from the details of the teeth. Finish off the sectional elevation, by showing the fitted bolts holding the assembly together.

The circles forming the end view are drawn in, as far as possible simultaneously with the elevation, as time can be saved by using the same compass setting. In the sectional half of the end view, the lightening holes in the cone will be seen as half ellipses, as the section plane cuts these through the centre. The drawing of the hexagon nuts in the end view is time consuming, so it would be sufficient to show a few of these, indicating the others by their centre lines as shown in the convention on page 9. For the method of showing the approximate profile of an involute tooth refer to page 18, and the completed drawing is on page 150.

For the completed drawing refer to page 150

TURBINE
MAIN GEAR WHEEL

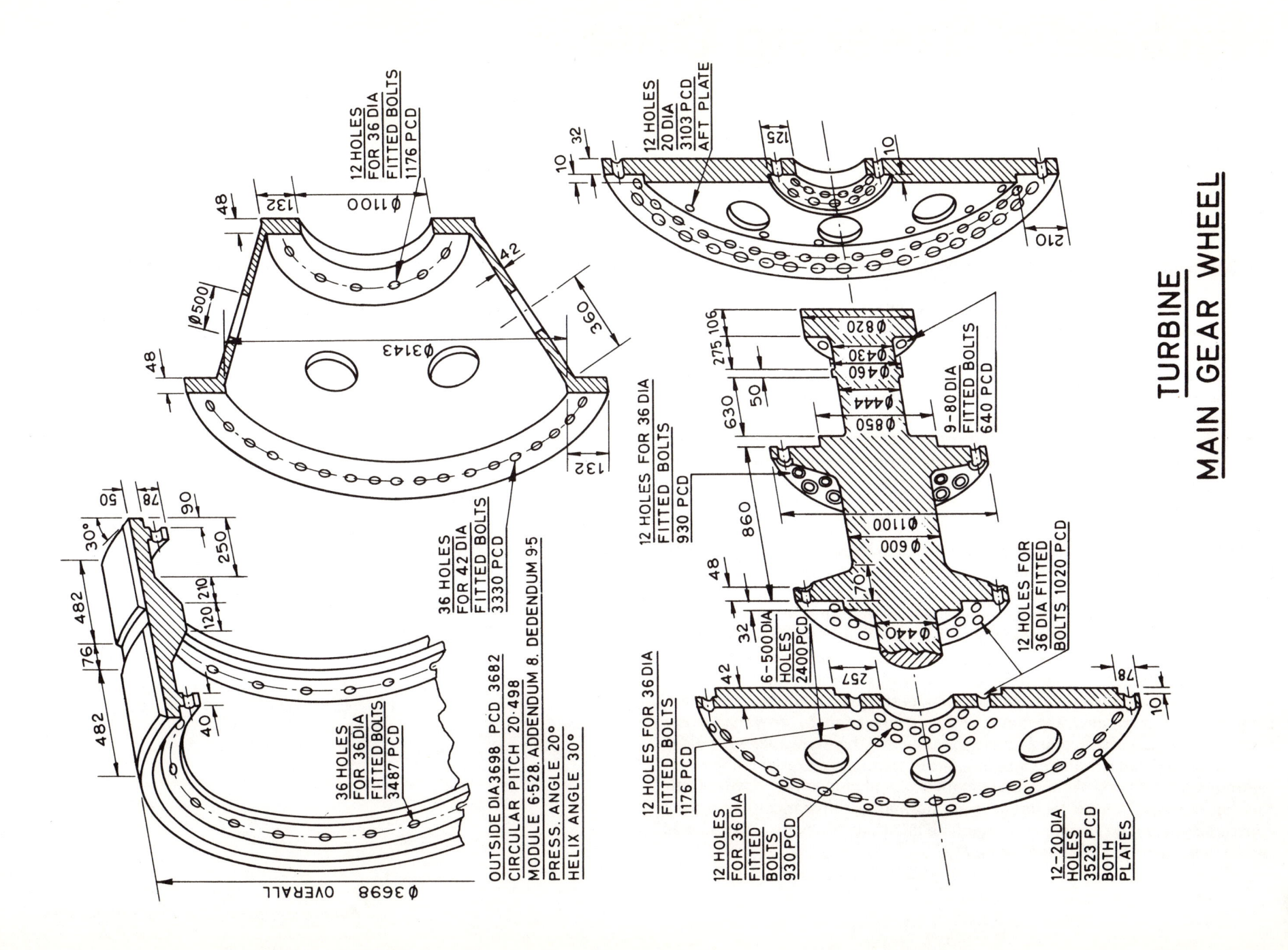

101

Hydraulic Steering Gear

DRAW

(i) Front elevation of the gear assembled, one half in section through centre of ram. (Assume tiller removed.)
(ii) Plan view, one half in section through centre of ram.

ALLOW 16mm FOR RUDDER WEAR DOWN, AND INDICATE THIS ON THE DRAWING. INCLUDE ON THE DRAWING A LIST OF MATERIALS.

Details are given of two rams, a trunnion and tiller, for a four ram hydraulic steering gear. Only two rams need to be shown, as the other two are similar. Before commencing to draw the required views, some thought should be given to the instruction regarding rudder wear down. It will be seen from the pictorial view that there is 360 between the forks of the ram to accommodate a trunnion, which is 336 deep at this point. This gives a total clearance of 24, and if the centre lines of the rams and trunnion coincided, this would mean an equal clearance of 12 top and bottom. To obtain the 16 required on the bottom of the trunnion boss however, the centre line of this component will have to be 4 above the centre line of the rams.

Decide on say the right hand side of both views to be in section, and draw in horizontal centre lines for the rams in both views. A vertical centre line in elevation is now drawn, and produced to the plan view underneath. From the dimensions given in the pictorial view, draw in the rams in elevation, the forked ends of which are held together by four 30 dia. fitted bolts, two of which will be partly seen in this view. In plan view, the rams could now be drawn in, together with the tiller, the centre of which is 504 from the centre line of the rams. The tiller arm slides in a bush 216 long, part of which will be seen in the section of the plan view. Continue with the elevation by drawing in the trunnion with its bushes, and as pointed out above the allowance for rudder wear down will be 16, if the centre line of the trunnion is drawn 4 above the centre line of the rams. Although allowance is made for a wear down of 16, this figure would not be required under normal circumstances, as a well lubricated rudder carrier bearing of the type illustrated on page 133, reduces rudder wear down to a very small amount.

Finish off both views by drawing in the guides, which are secured to the rams by four 25 dia. fitted bolts, the oil reservoir on the top of the trunnion, the oil holes and grooves, and the syphon tube.

For the completed drawing refer to page 151

HYDRAULIC STEERING GEAR

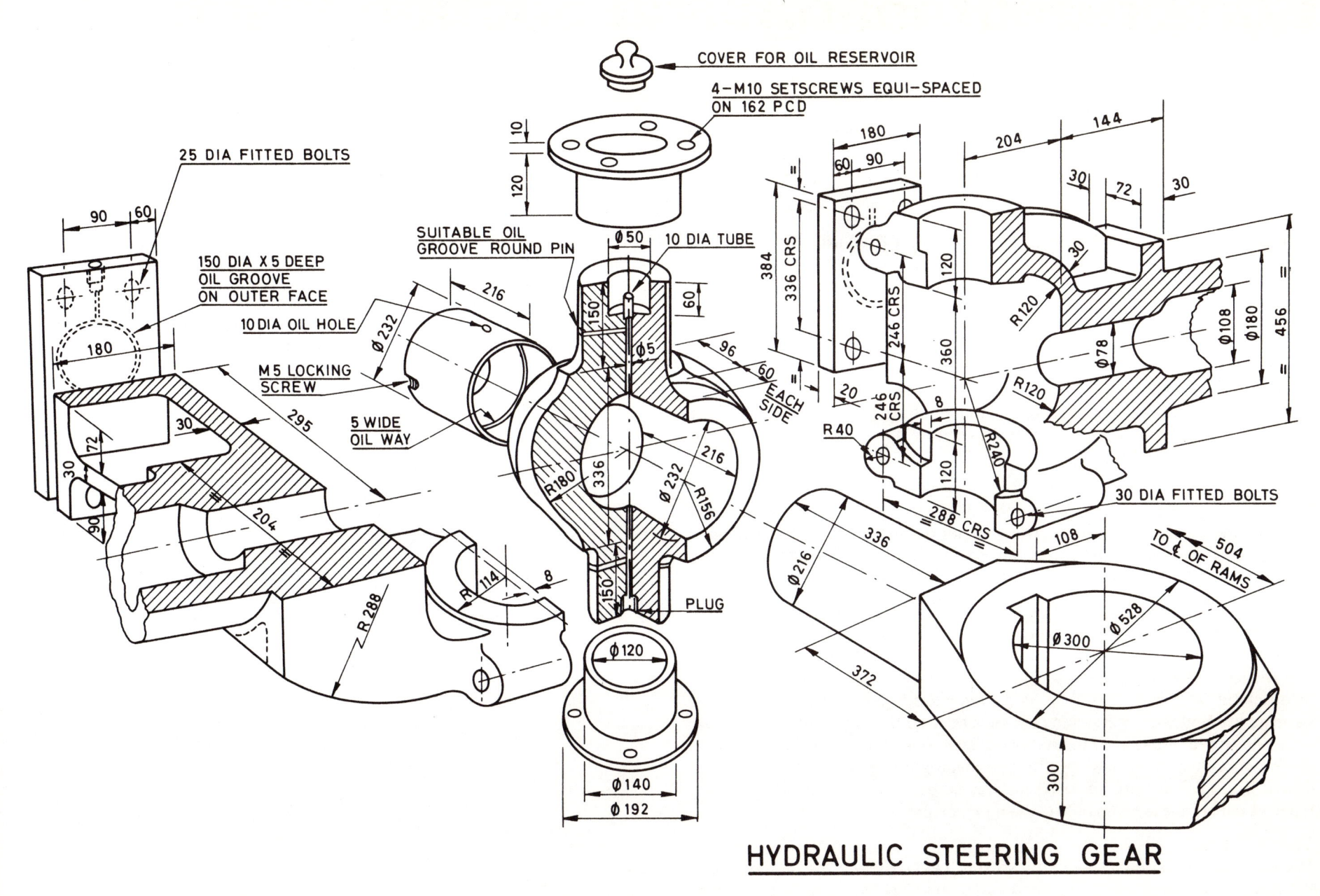

Compressor Piston and Suction Valve

DRAW

(i) Sectional elevation through axis of gudgeon pin showing all parts assembled.
(ii) End elevation.
(iii) Plan view with valve assembly removed, one half to be an outside view, one half a section 70 above gudgeon pin centre line.

INDICATE VALVE LIFT, AND INCLUDE ON THE DRAWING A LIST OF MATERIALS.

The piston and suction valve illustrated on opposite page is from a Freon 12 compressor, where the suction port is in the liner. On the up stroke of the piston, with suction valve closed, gas enters the piston in the space below the valve, and on the downstroke the suction valve lifts, allowing this gas to enter the space above the piston, where it is compressed on the upstroke. This arrangement isolates the suction side of the machine from the crankcase, as there are two piston rings below the gudgeon pin. The piston is splash lubricated from the sump.

For students who are not familiar with this suction valve arrangement, the pictorial views of the valve seat, valve, and guard show their relative positions on assembly. The collar stud and nut secure the whole assembly to the piston.

The sectional elevation of the piston should not present any difficulty, as the pictorial view of this provides ample information. The valve seat of 81 dia. fits in a recess of the same diameter, in the piston head, and the valve ring of external dia. 74.5 and bore 52.5, is sandwiched between the seat and the valve guard. Notice an M6 locking screw for the collar stud, and also that two holes in the seat are tapped to enable the seat to be withdrawn. The gudgeon pin is not shown pictorially, but its dimensions are given so that it can be drawn in, together with its M10 locking screw. The end of this screw is tapered, to fit into a tapered hole in the gudgeon pin.

The plan view will show three of the six ribs, cast between the stud boss and the wall of the piston, whilst the sectional plan will show the piston cut away 45° each side of the gudgeon pin centre line.

An inverted plan view is included on the completed drawing on page 152, but from a point of view of time this would not be expected in the examination. It might well be asked for however, instead of one of the other views.

For the completed drawing refer to page 152

COMPRESSOR PISTON AND SUCTION VALVE

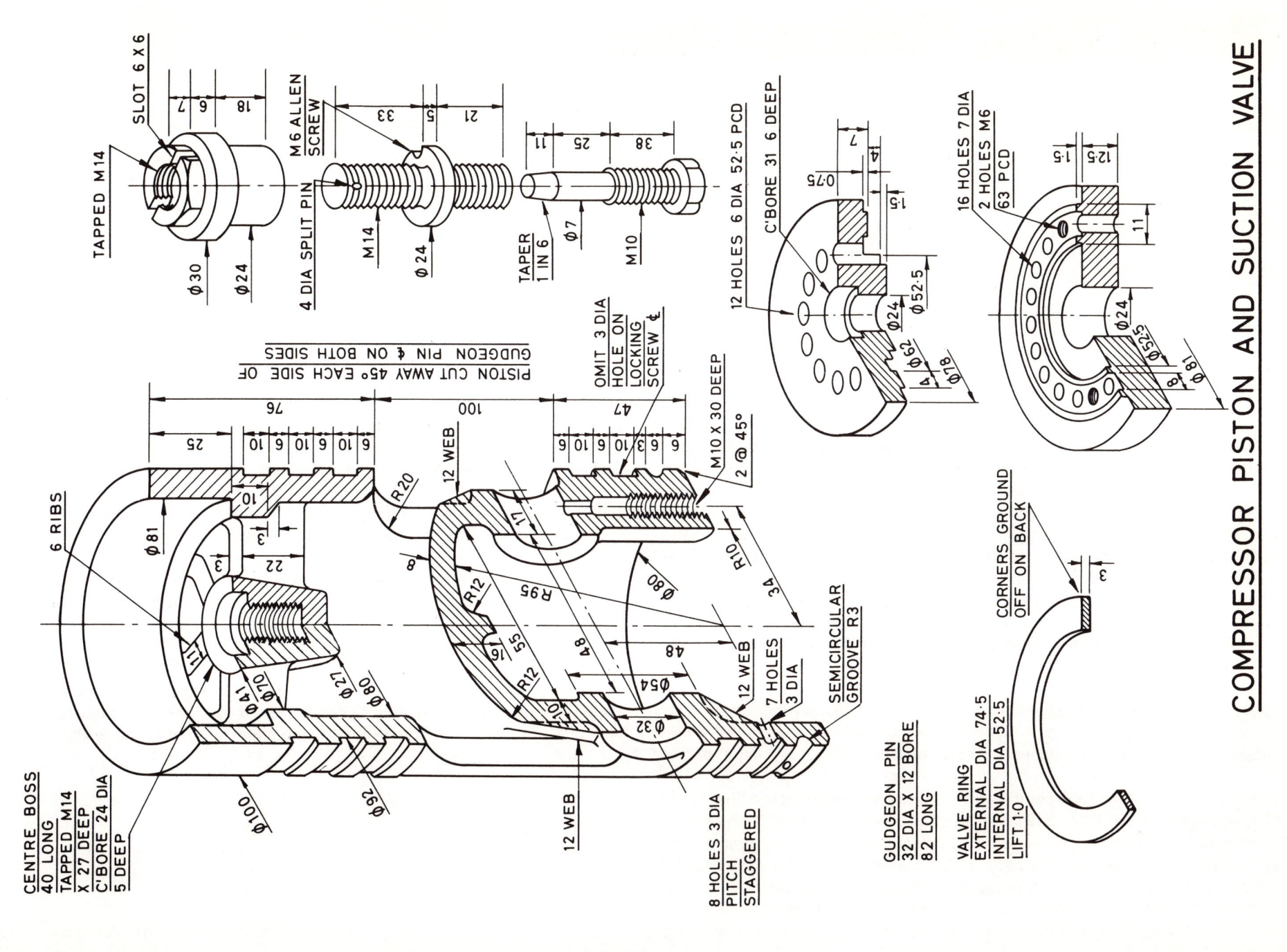

Plate Type Gauge Glass

DRAW

(i) Sectional elevation through the centre lines of both glasses, showing gauge assembled. Show steam cock in section, and an outside view of water and drain cocks.
(ii) End elevation.
(iii) Sectional plan through steam header and steam cock.

INCLUDE ON THE DRAWING A LIST OF MATERIALS.

The gauge glass illustrated pictorially on the opposite page is a common type in use on high pressure water tube boilers, and candidates preparing for a Steam Certificate should not find difficulty in assembling the components. For those who are not familiar with its construction, however, it functions briefly as follows:—

Steam and water are admitted to their respective headers by parallel bore shut off cocks, of the sleeve packed type, which are mounted on the steam drum. From the water header, the water rises through a 20 dia. ball valve to the space between the glasses, and the level can be easily seen, as there is illumination behind the gauge glass. The louvre plate is fitted to deflect the light, and hence give the illusion of an illuminated bubble on the surface of the water. Plugs are provided in both headers for cleaning the glass, and the steam and water ports. It will be seen from the pictorial view of one glass, that there is a thin sheet of mica sandwiched between the C.A.F. joint and the glass. This is fitted to protect the glass from the chemicals introduced into the boiler feed water.

Commence the sectional elevation, by drawing in the vertical centre line through top and bottom headers. A horizontal centre line through the steam header is now drawn, allowing enough space above this for the upper end of the 20 dia. tube, together with the 15 deep square on the top cleaning plug. Draw in the steam header from the dimensions given, and locate this on the body of the steam cock by its 38 dia. spigot. Draw in the tube in the steam header, together with its lantern ring, packing ring and gland, keeping the ports in the tube in line with the ports in the header. The 30 dia. collar on the tube, fits into the 1 BSP hole in the top of the gauge centre piece, secured by a gland nut. Locate both glasses on the centre piece, and then draw in gauge glass body, louvre plate and back plate. Show about four M12 nuts securing the back plate, and also about four M10 pinching screws. Finish off the sectional view by drawing in the water header, which is secured to the centre piece by a split gland nut screwed 1½ BSP. Locate the water cock on the 38 dia. spigot on the header, and draw this next. Note this cock is not required in section, nor is the drain cock, which screws into the bottom of it on a 1 BSP thread. The cock operating levers would finish this view.

The end view projected from the sectional elevation, and to some extent drawn simultaneously with it, could now be finished off, by drawing in the width of the various components seen in this view from the dimensions given in the pictorial views of the various components.

The sectional plan through the steam header and cock, will show two M16 studs securing header to cock body, and will also show details of the cock plug, packing and sleeve. Note that these items are similar for all three cocks.

On the completed drawing on page 153, a section through the glasses is also included, but from a point of view of the time allowed for the question this would not be required. It might well be asked for however, instead of the sectional plan required in part (iii). Note the materials in this gauge glass.

For the completed drawing refer to page 153

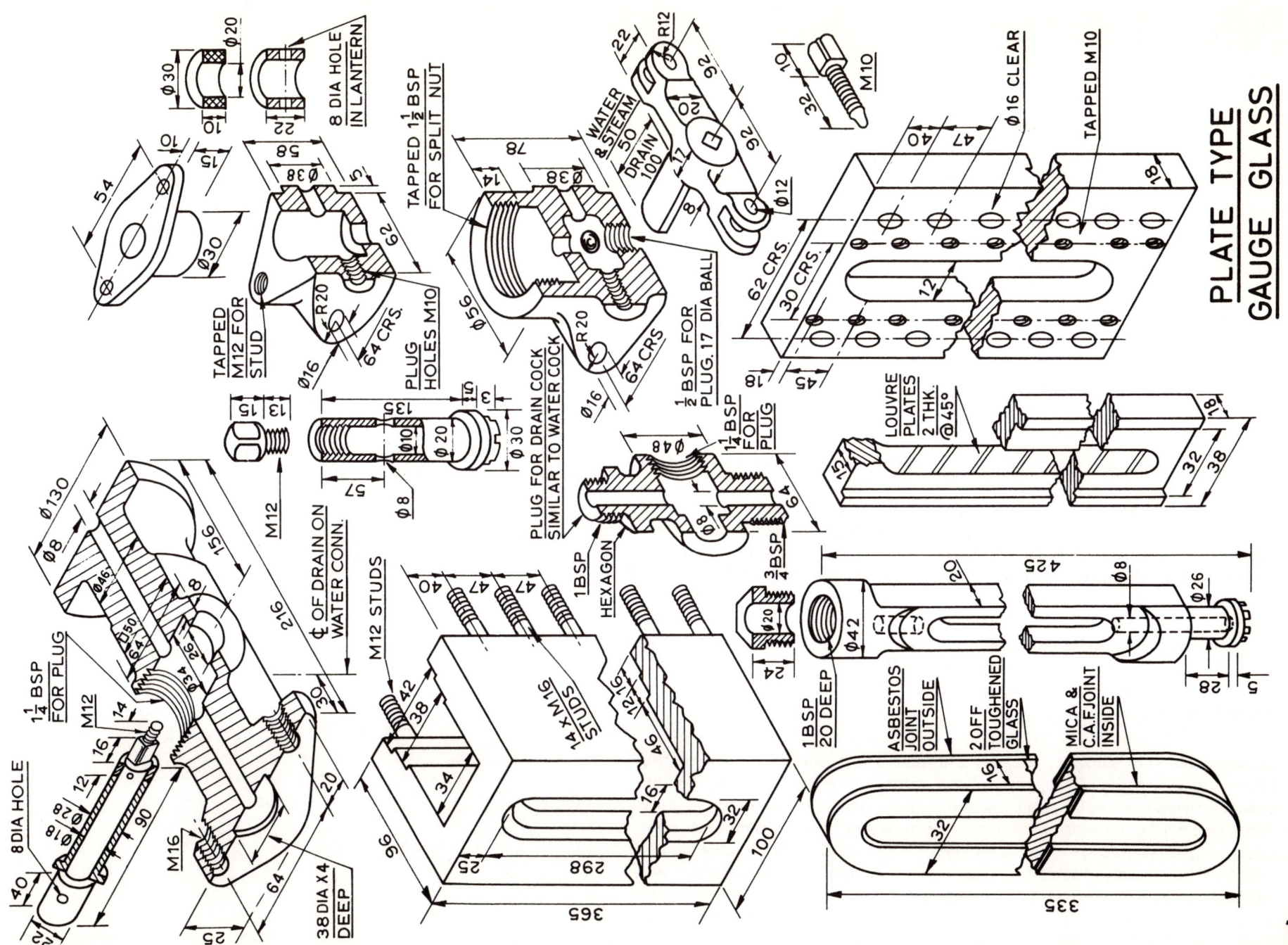

PLATE TYPE GAUGE GLASS

107

Mechanical Lubricator

DRAW

(i) Front elevation with all parts assembled.
(ii) Sectional end elevation, the section plane being taken through a pump.

INCLUDE ON THE DRAWING A LIST OF MATERIALS.

Engineers whose sea service has been only on turbine ships may not be familiar with the mechanical lubricator illustrated on the opposite page, but with a few exceptions due to limitations of space, the pictorial view is an exploded view so that little difficulty should be experienced in locating the various components.

The unit illustrated is a two pump model, but the number of pumps would, of course, depend on the number of points to be lubricated. These pumps are operated by eccentrics mounted on a shaft which passes through the container, and running in bearings screwed into the ends of the container. The worm and wheel drive for this shaft has been omitted. On the suction stroke of the differential pump plunger, oil is drawn from the container through the 5 dia. port, the four 3 dia. holes in suction valve seat, and the ball suction valves to the space below the plunger. On the delivery stroke of the plunger the oil is forced through the delivery valves and nozzle into the sight glass, which is filled with water. When the globule of oil is sufficiently large, it travels up the guide wire, and into the delivery pipe above the 6 dia. non return ball valve.

Draw in the front elevation of the container and cover, and mark off the horizontal centre line of the pumps. Project this centre line across to the end view, and draw in container and cover in section. On the front elevation, draw in the vertical centre lines of the pumps 48 each side of container centre line. Each pump is secured to the container by 4 x M6 screws in a 37 square flange. Finish off the pumps and flushing triggers in front elevation, and draw in the two sight glasses and holders, which pass through the 14 thick projection in front of the container. A serrated nut above and below this projection compresses the cork gaskets, keeping the unit oil tight.

Proceed to the end view, and draw a pump in section from the details given, each component taking the relative position shown. To locate the plunger at the end of its delivery stroke, the thumb nut is drawn and the 15 mean dia. spring drawn at its free length of 15. Note that when the suction valve seat is screwed into pump body, sufficient thread remains for a ½ BSP cap nut to be fitted. Finish off this view by drawing in the operating eccentric at the end of the delivery stroke, and the 16 dia. spindle in section.

Notice on the pictorial view that there is a ¾ BSP hole in the projection cast on the front, and part of one side of the container. This is for securing the contents sight glass, which could now be drawn in, together with its hexagon bar base, to finish the front elevation.

For the solution to this example refer to page 154

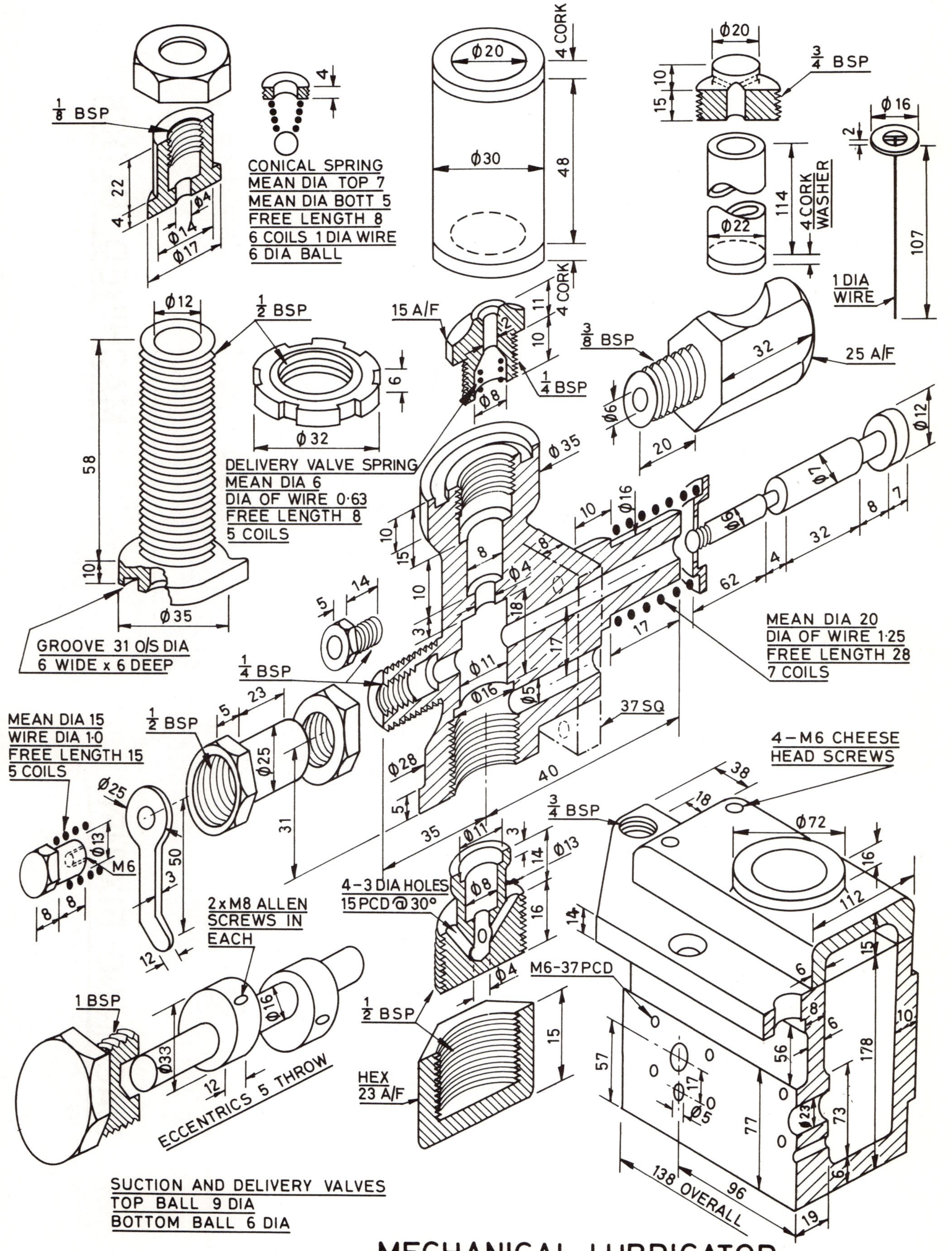

MECHANICAL LUBRICATOR

MACHINING FIXTURE

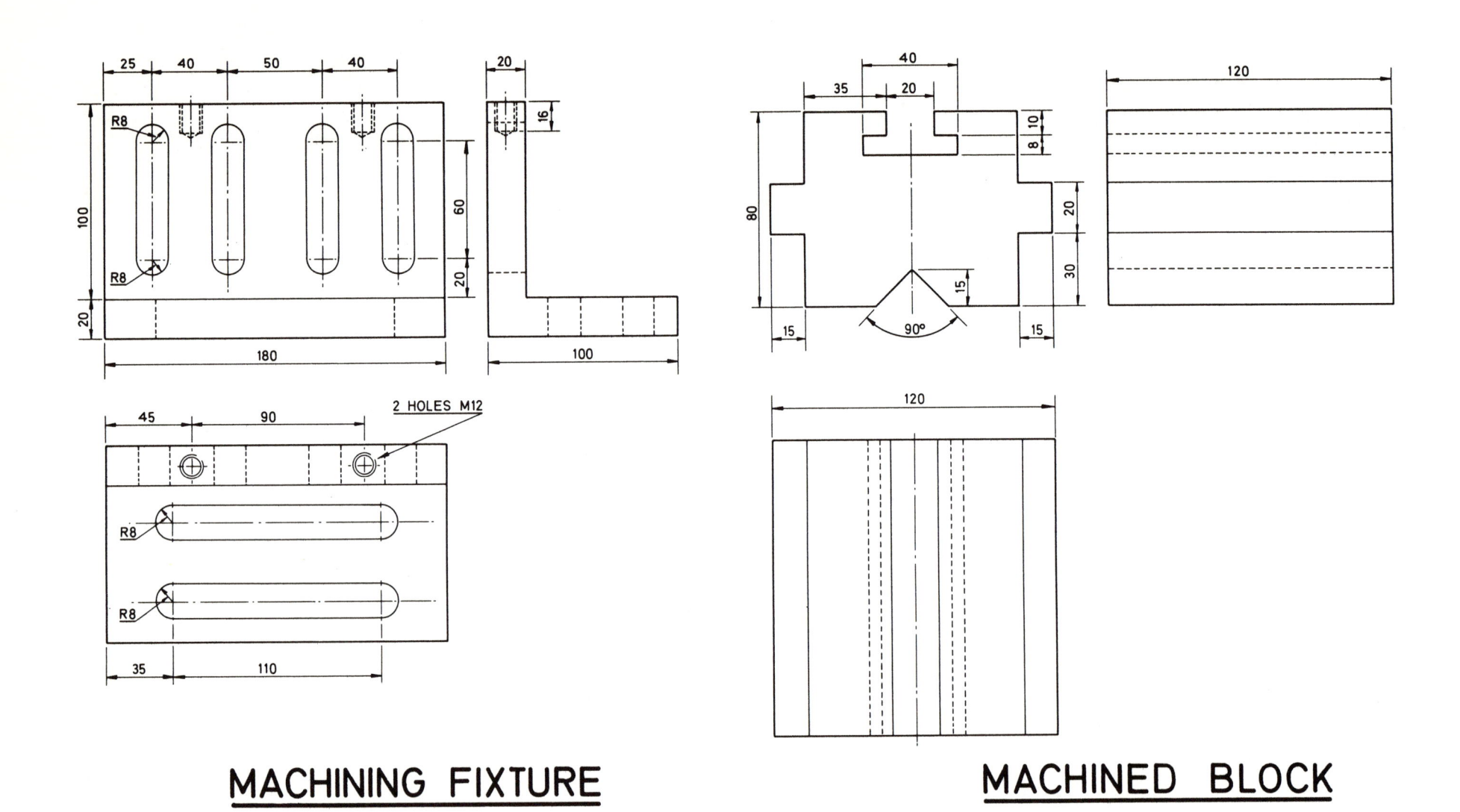

MACHINED BLOCK

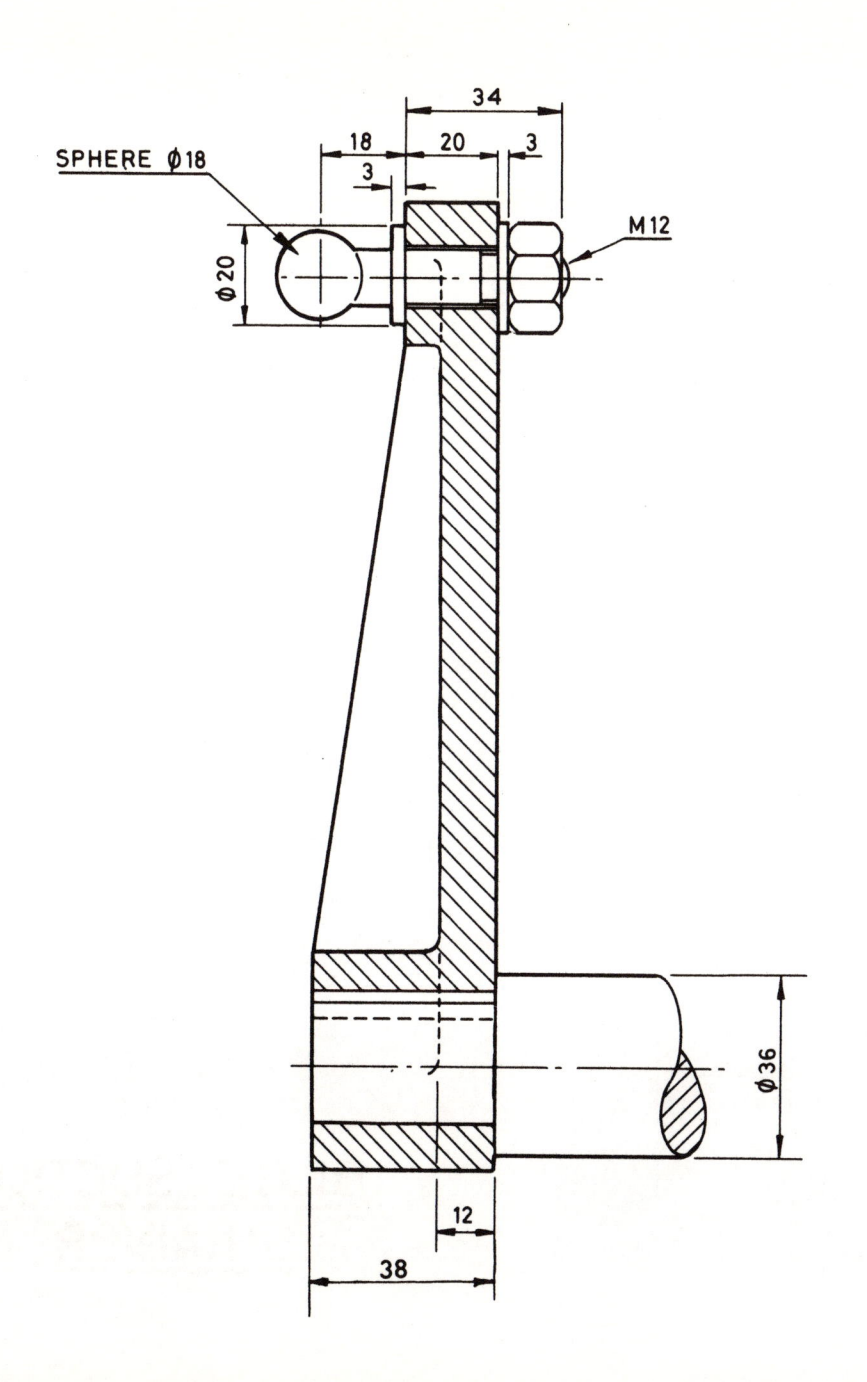

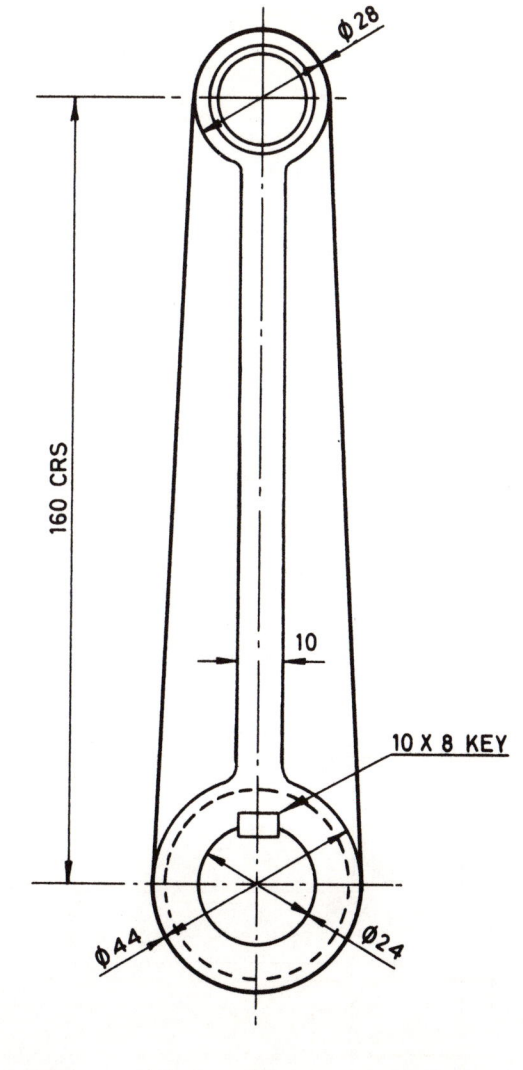

FUEL CONTROL LEVER

111

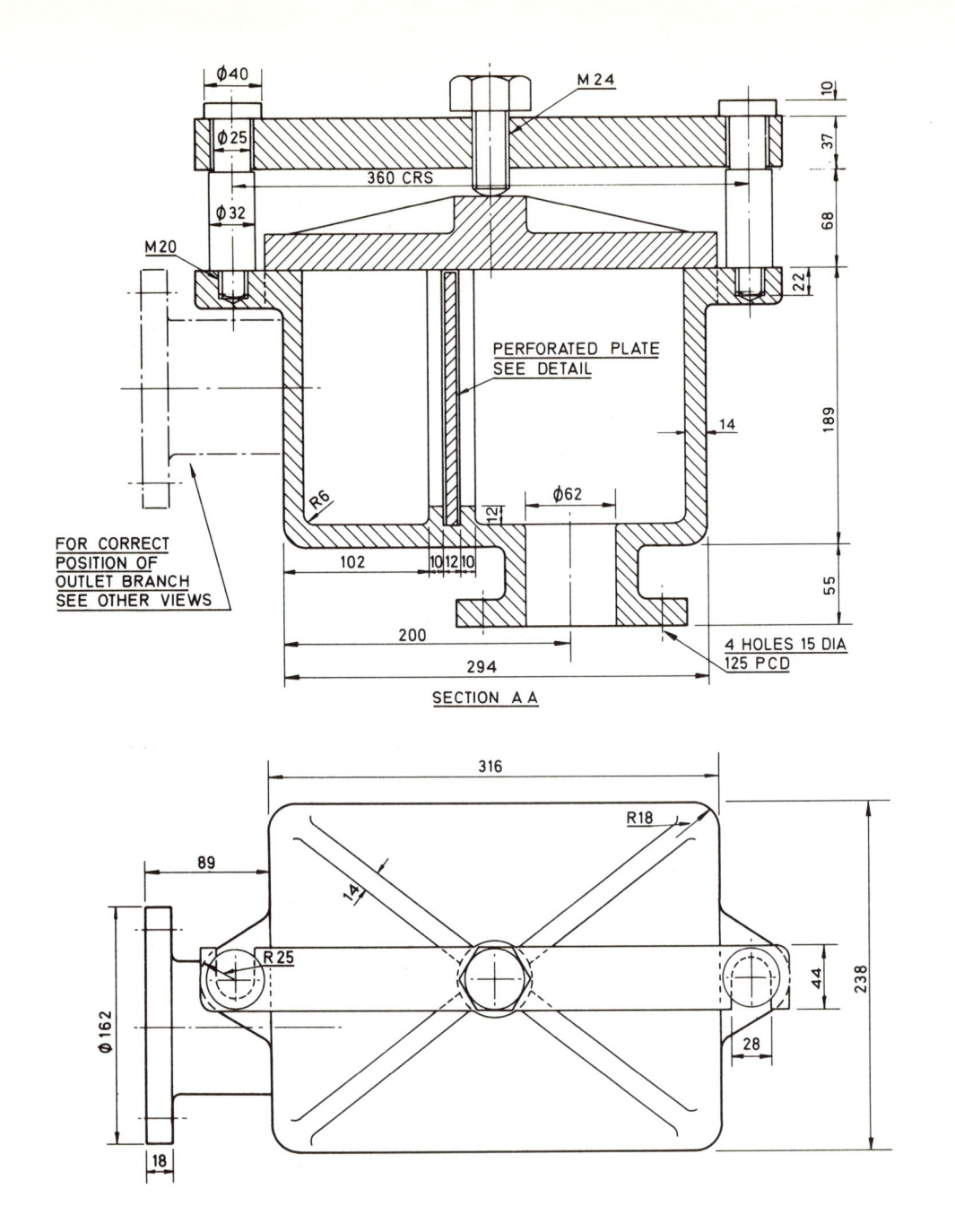

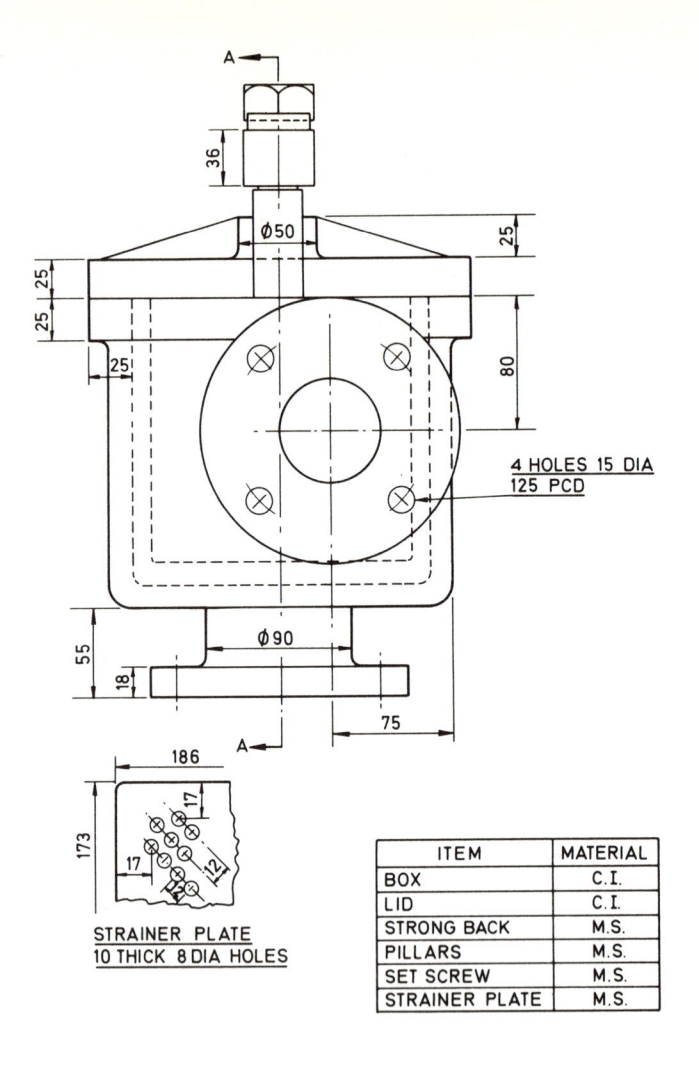

BILGE SUCTION STRAINER

ITEM	MATERIAL
BOX	C.I.
LID	C.I.
STRONG BACK	M.S.
PILLARS	M.S.
SET SCREW	M.S.
STRAINER PLATE	M.S.

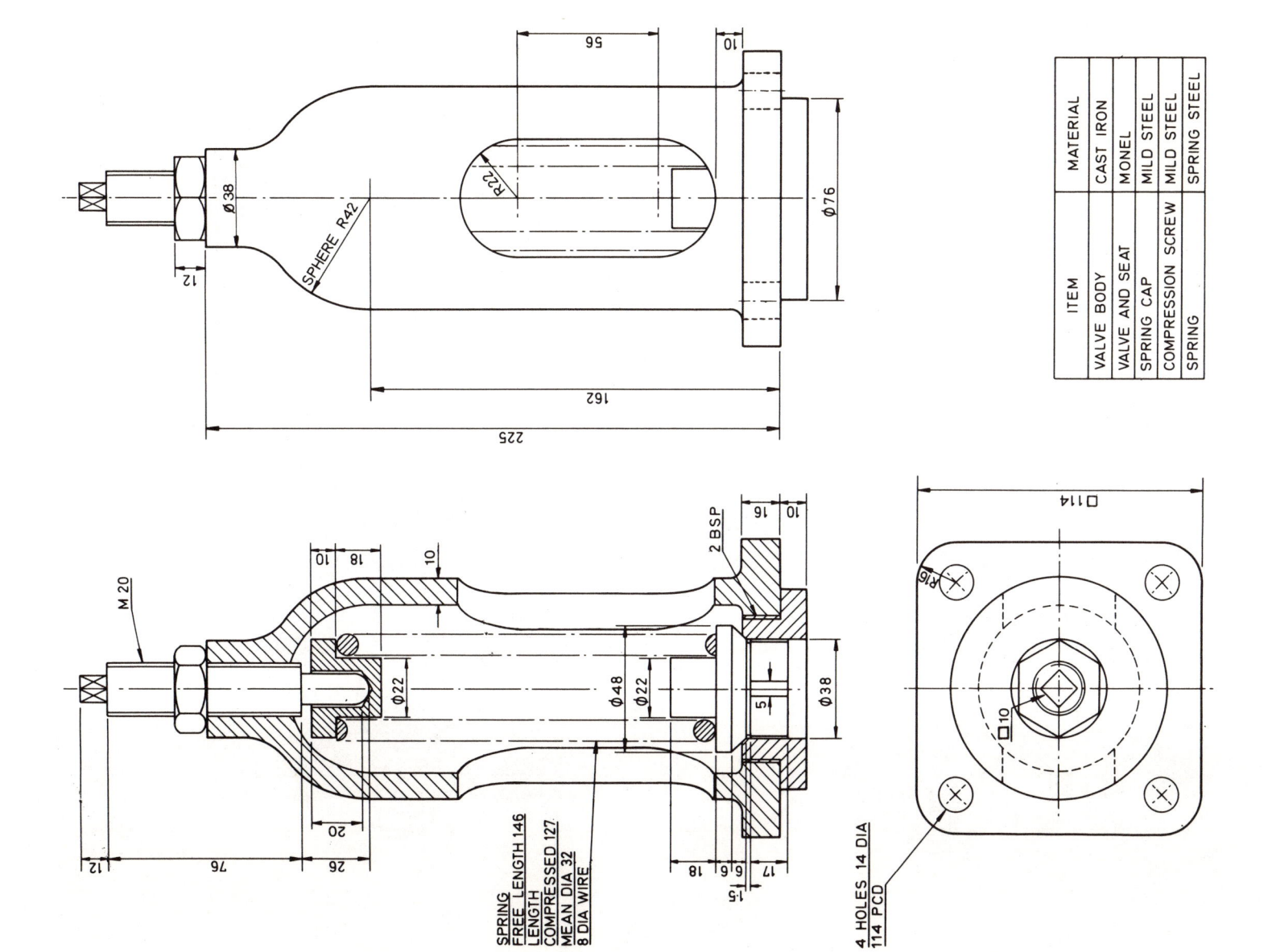

CYLINDER RELIEF VALVE

ITEM	MATERIAL
VALVE BODY	CAST IRON
VALVE AND SEAT	MONEL
SPRING CAP	MILD STEEL
COMPRESSION SCREW	MILD STEEL
SPRING	SPRING STEEL

SPRING
FREE LENGTH 146
COMPRESSED LENGTH 127
MEAN DIA 32
8 DIA WIRE

4 HOLES 14 DIA
114 PCD

113

CRANE HOOK

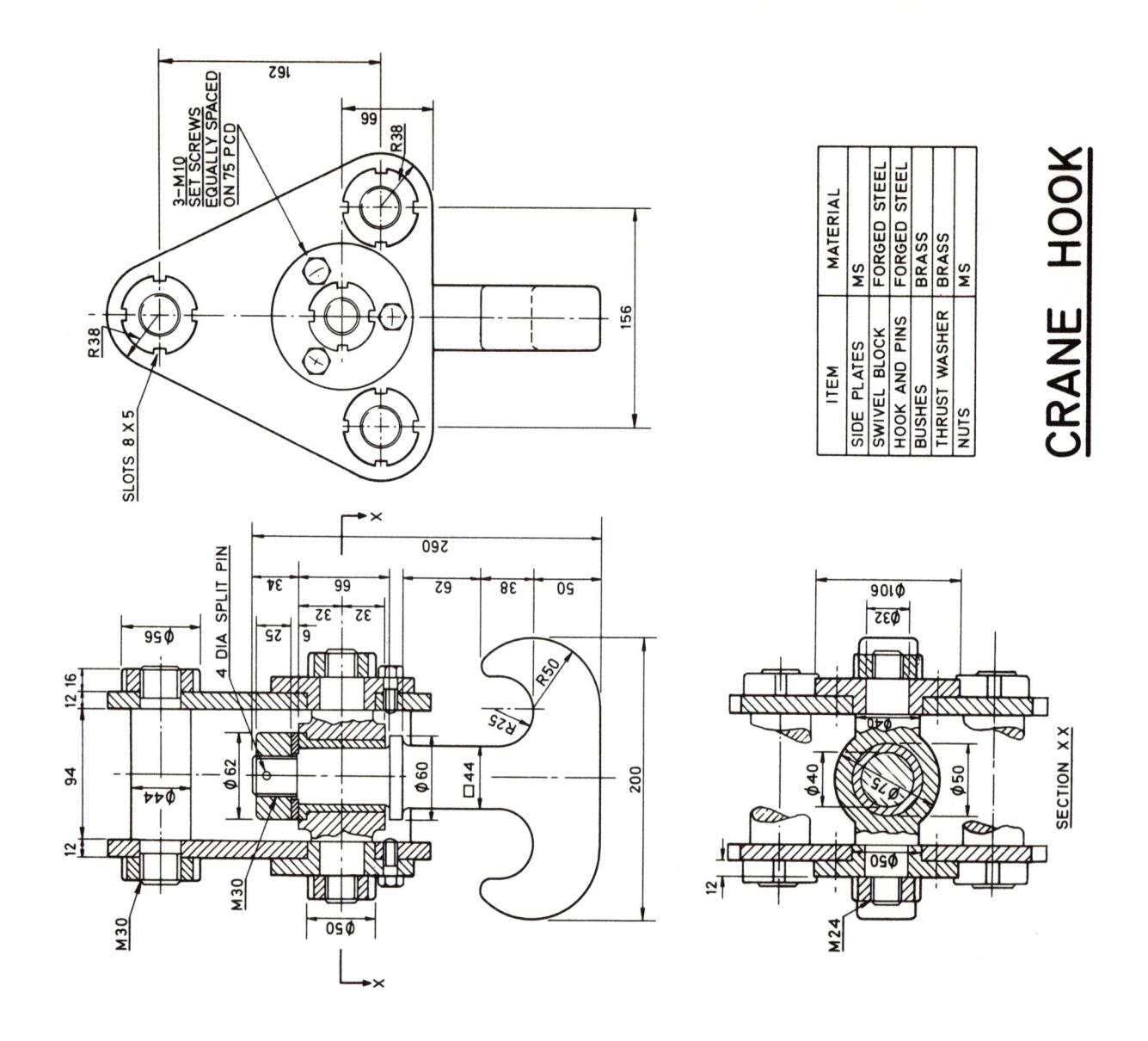

ITEM	MATERIAL
SIDE PLATES	MS
SWIVEL BLOCK	FORGED STEEL
HOOK AND PINS	FORGED STEEL
BUSHES	BRASS
THRUST WASHER	BRASS
NUTS	MS

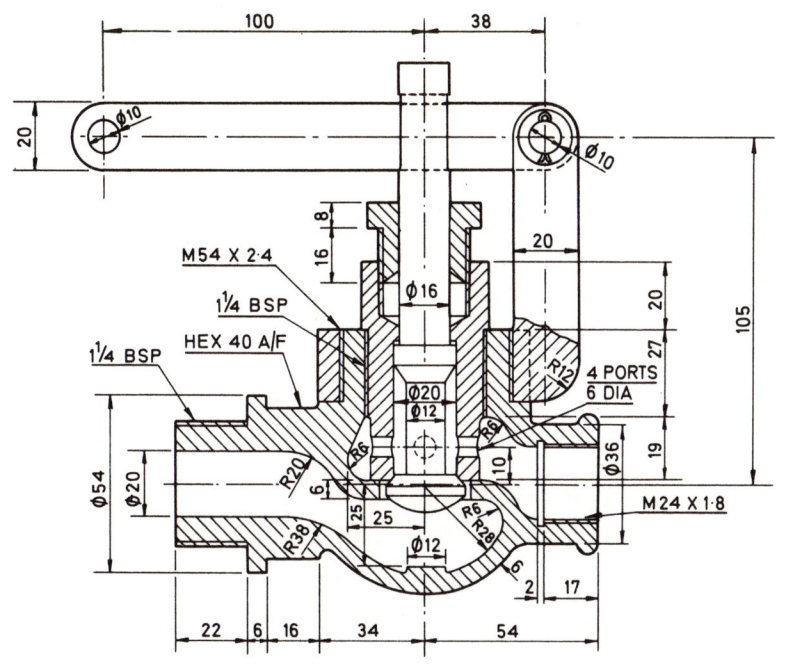

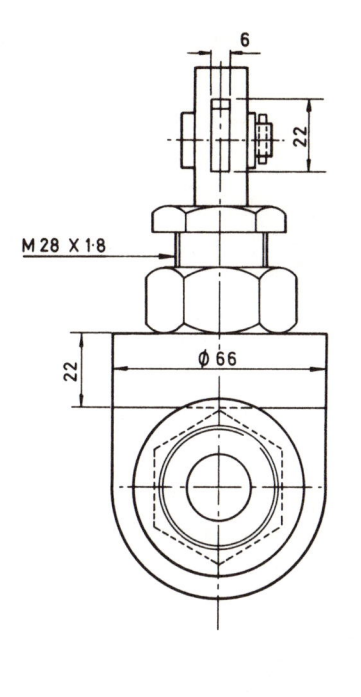

ITEM	MATERIAL
VALVE BODY	GM
VALVE SPINDLE	STAINLESS STEEL
VALVE SEAT	MONEL
GLAND	GM
FULCRUM NUT	GM
OPERATING LEVER	MS
FULCRUM PIN	MS

CONTROL VALVE

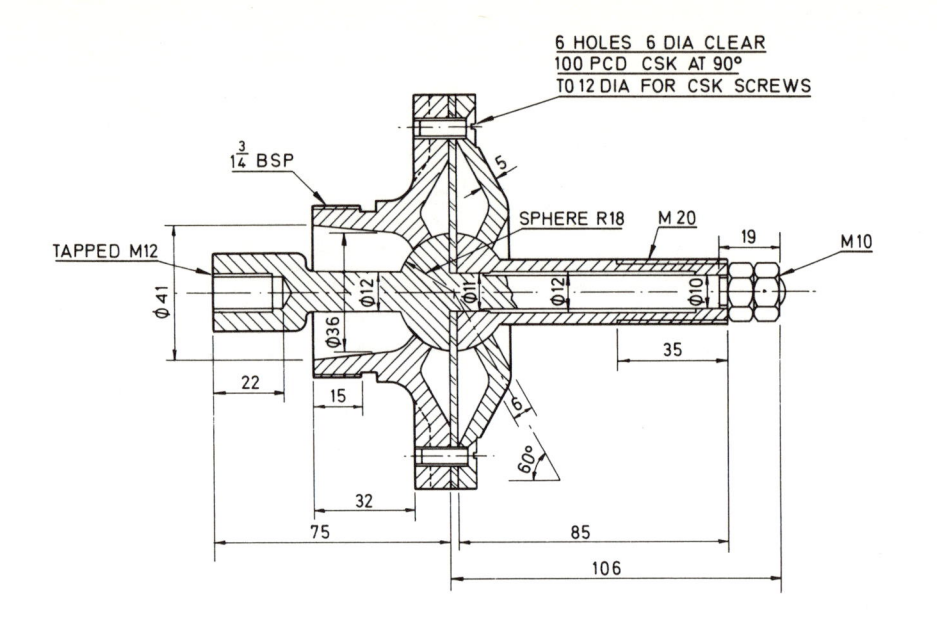

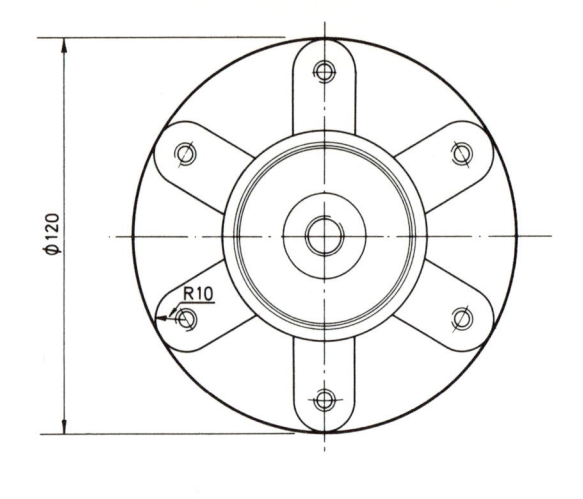

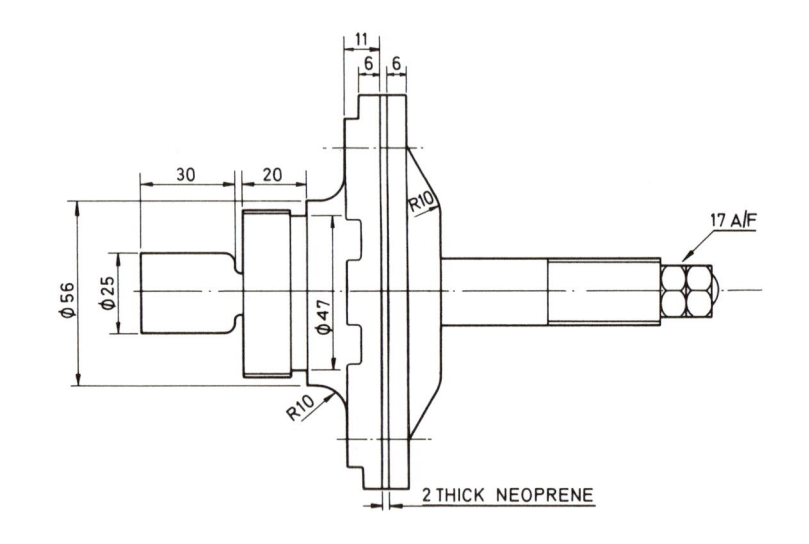

SEALED BALL JOINT

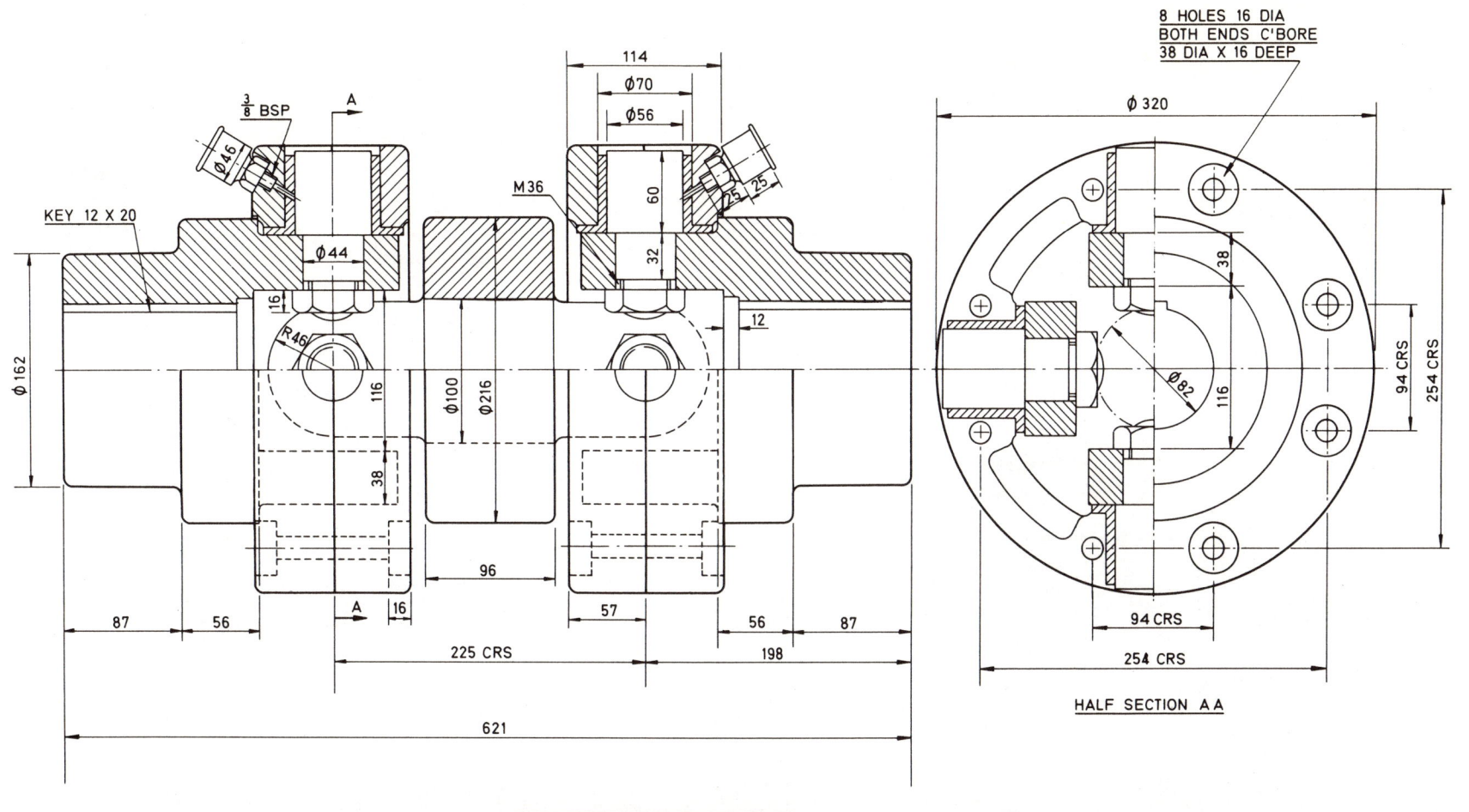

ITEM	MATERIAL
COUPLING BODY	CAST STEEL
FORKED ENDS	FORGED STEEL
FORKED LINK	FORGED STEEL
PINS	M.S.
BUSHES	BRASS
TIE BOLTS	M.S.

UNIVERSAL COUPLING

117

OIL FUEL STRAINER

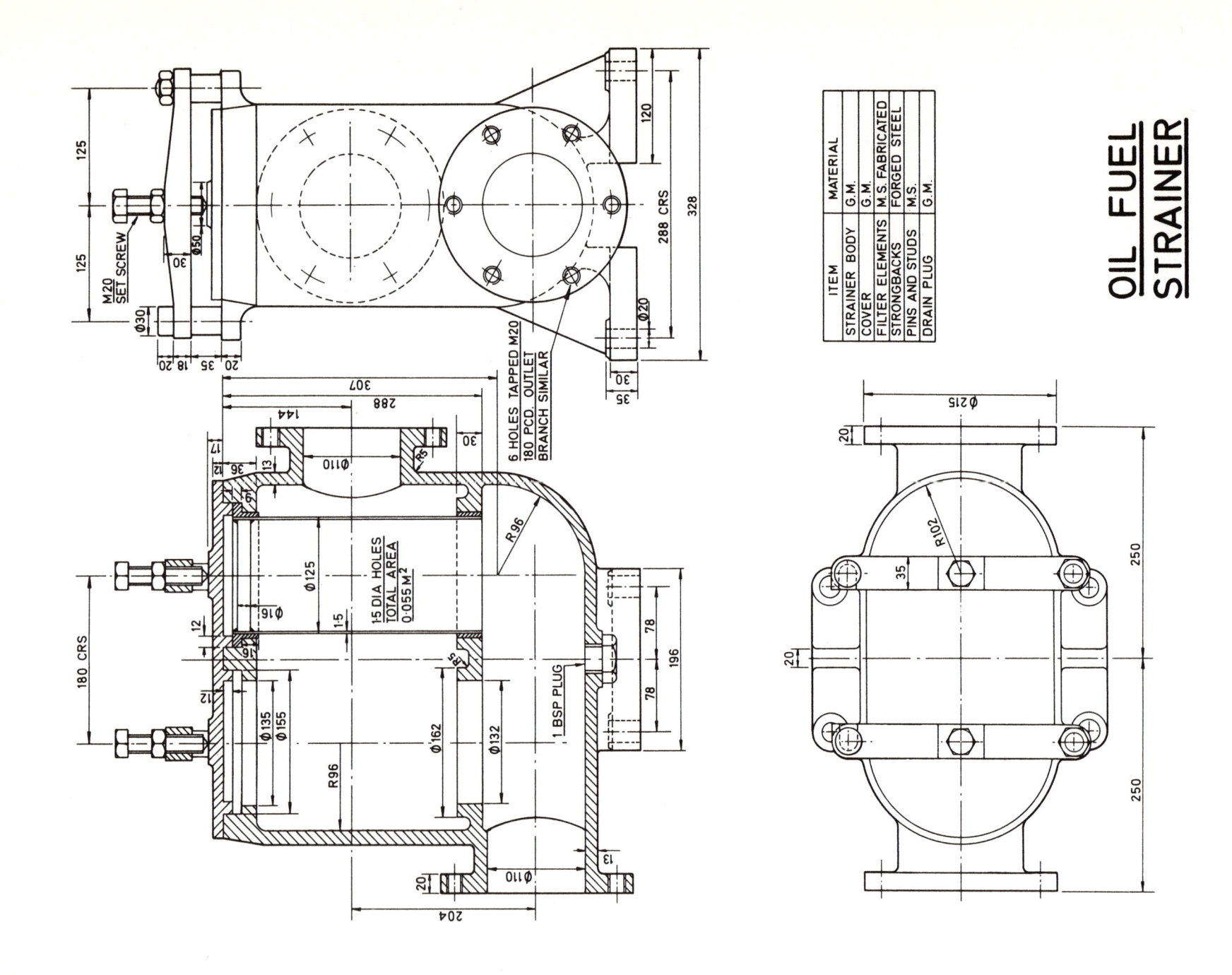

ITEM	MATERIAL
STRAINER BODY	G.M.
COVER	G.M.
FILTER ELEMENTS	M.S. FABRICATED
STRONGBACKS	FORGED STEEL
PINS AND STUDS	M.S.
DRAIN PLUG	G.M.

PARALLEL SLIDE STOP VALVE

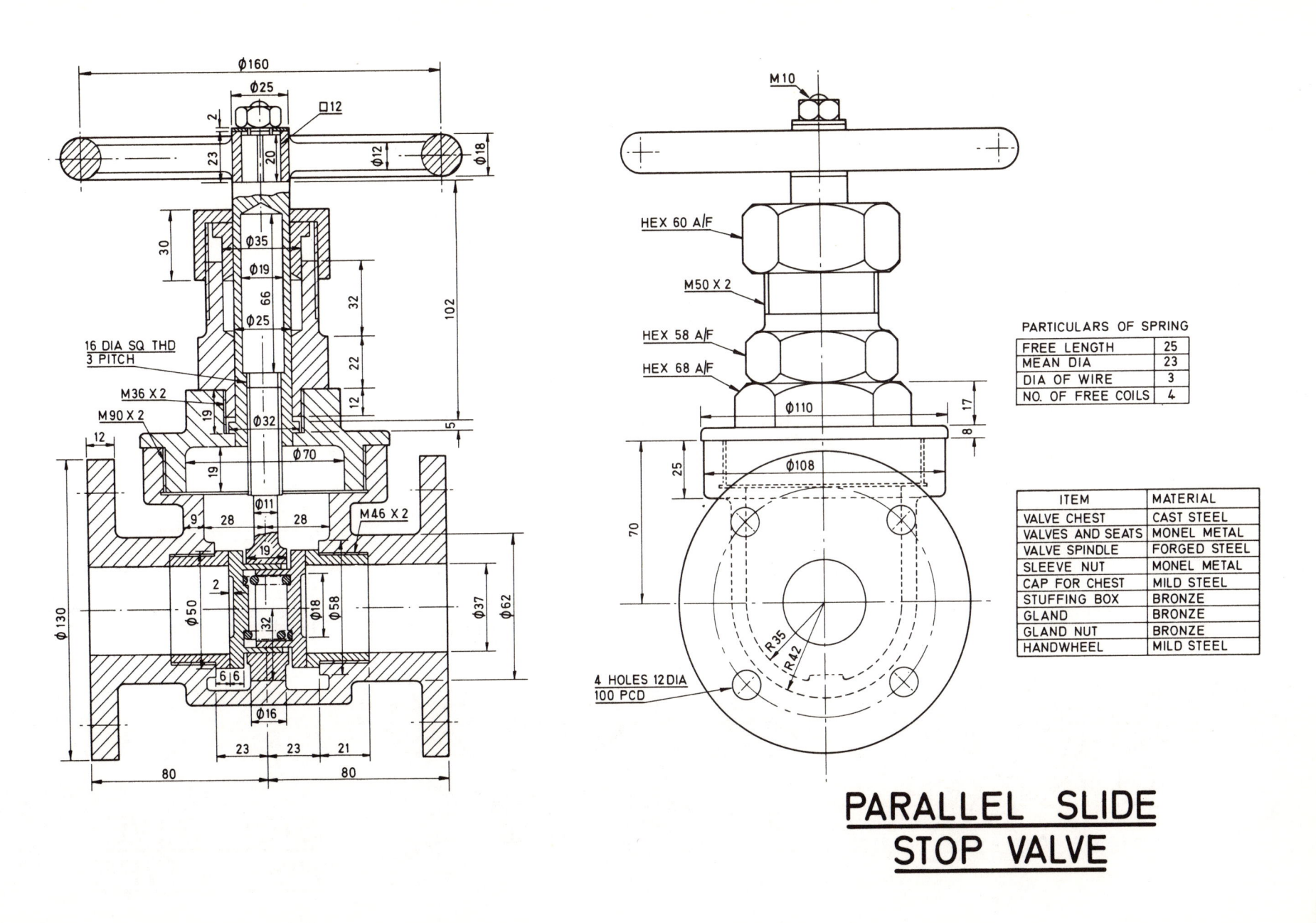

PARTICULARS OF SPRING

FREE LENGTH	25
MEAN DIA	23
DIA OF WIRE	3
NO. OF FREE COILS	4

ITEM	MATERIAL
VALVE CHEST	CAST STEEL
VALVES AND SEATS	MONEL METAL
VALVE SPINDLE	FORGED STEEL
SLEEVE NUT	MONEL METAL
CAP FOR CHEST	MILD STEEL
STUFFING BOX	BRONZE
GLAND	BRONZE
GLAND NUT	BRONZE
HANDWHEEL	MILD STEEL

STOP VALVE (PISTON TYPE)

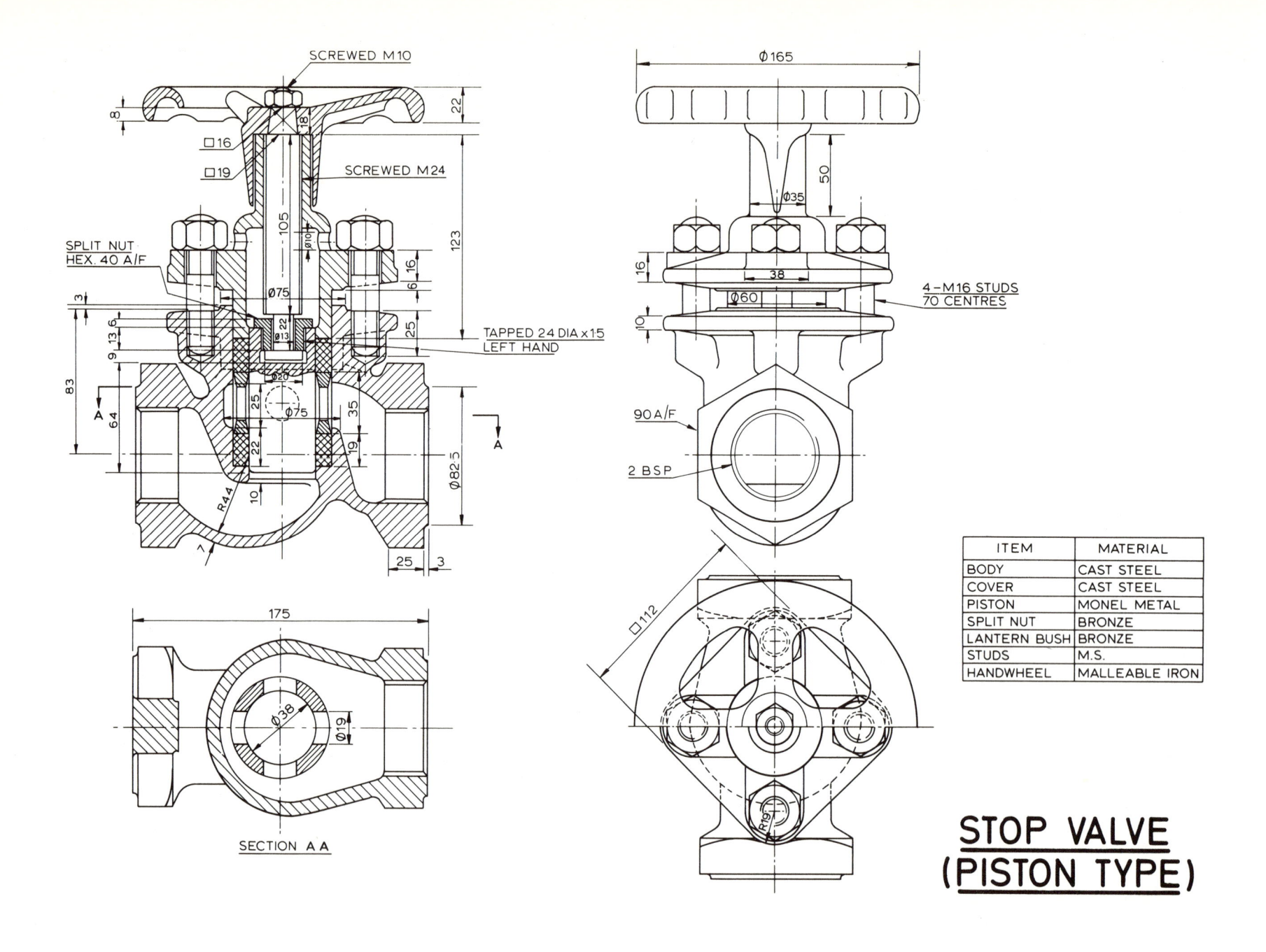

ITEM	MATERIAL
BODY	CAST STEEL
COVER	CAST STEEL
PISTON	MONEL METAL
SPLIT NUT	BRONZE
LANTERN BUSH	BRONZE
STUDS	M.S.
HANDWHEEL	MALLEABLE IRON

BALLAST VALVE CHEST FOR OIL OR WATER

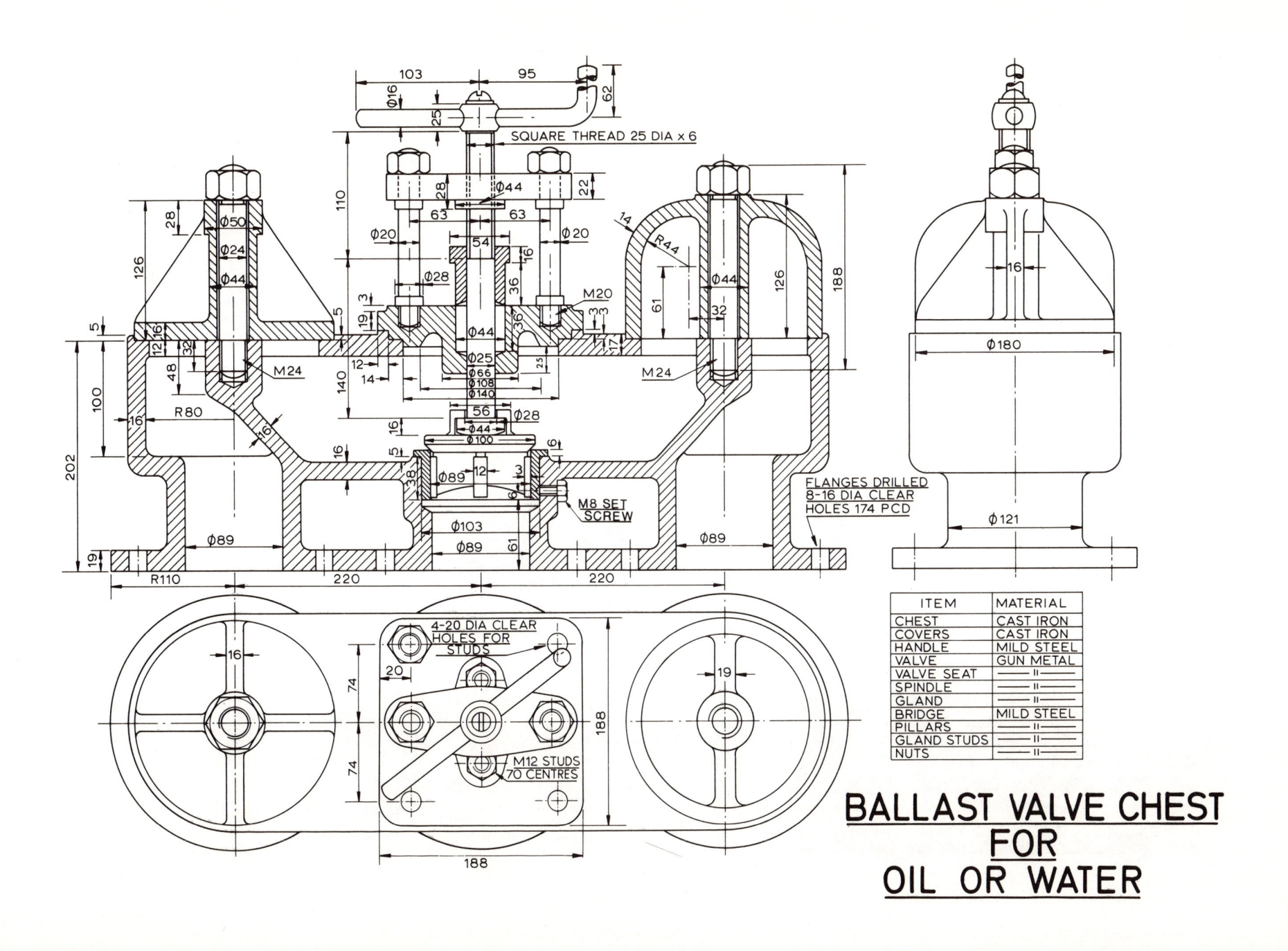

ITEM	MATERIAL
CHEST	CAST IRON
COVERS	CAST IRON
HANDLE	MILD STEEL
VALVE	GUN METAL
VALVE SEAT	—"—
SPINDLE	—"—
GLAND	—"—
BRIDGE	MILD STEEL
PILLARS	—"—
GLAND STUDS	—"—
NUTS	—"—

SQUARE THREAD 25 DIA x 6

FLANGES DRILLED 8-16 DIA CLEAR HOLES 174 PCD

M8 SET SCREW

4-20 DIA CLEAR HOLES FOR STUDS

M12 STUDS 70 CENTRES

121

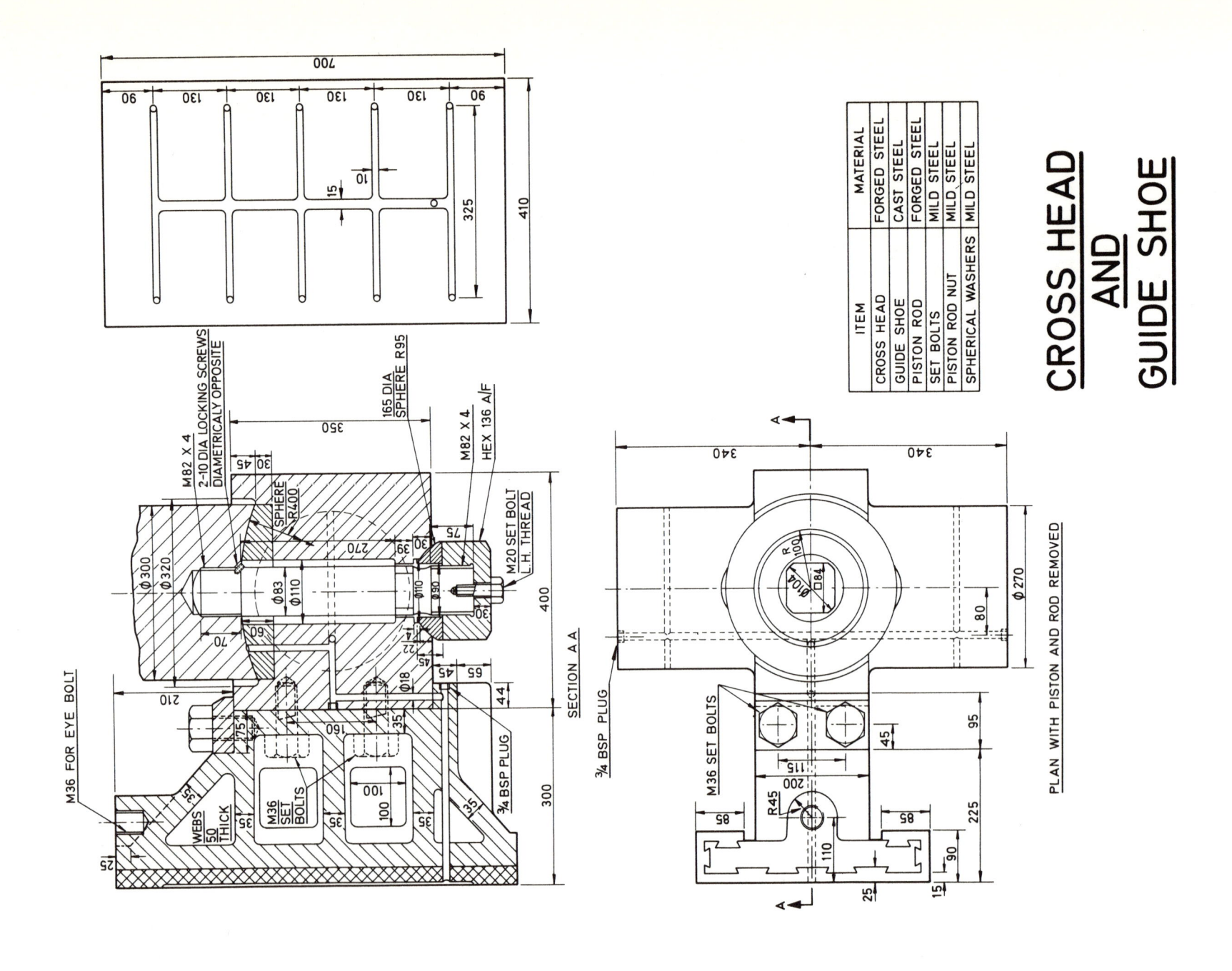

CROSS HEAD AND GUIDE SHOE

ITEM	MATERIAL
CROSS HEAD	FORGED STEEL
GUIDE SHOE	CAST STEEL
PISTON ROD	FORGED STEEL
SET BOLTS	MILD STEEL
PISTON ROD NUT	MILD STEEL
SPHERICAL WASHERS	MILD STEEL

FEED CHECK VALVE

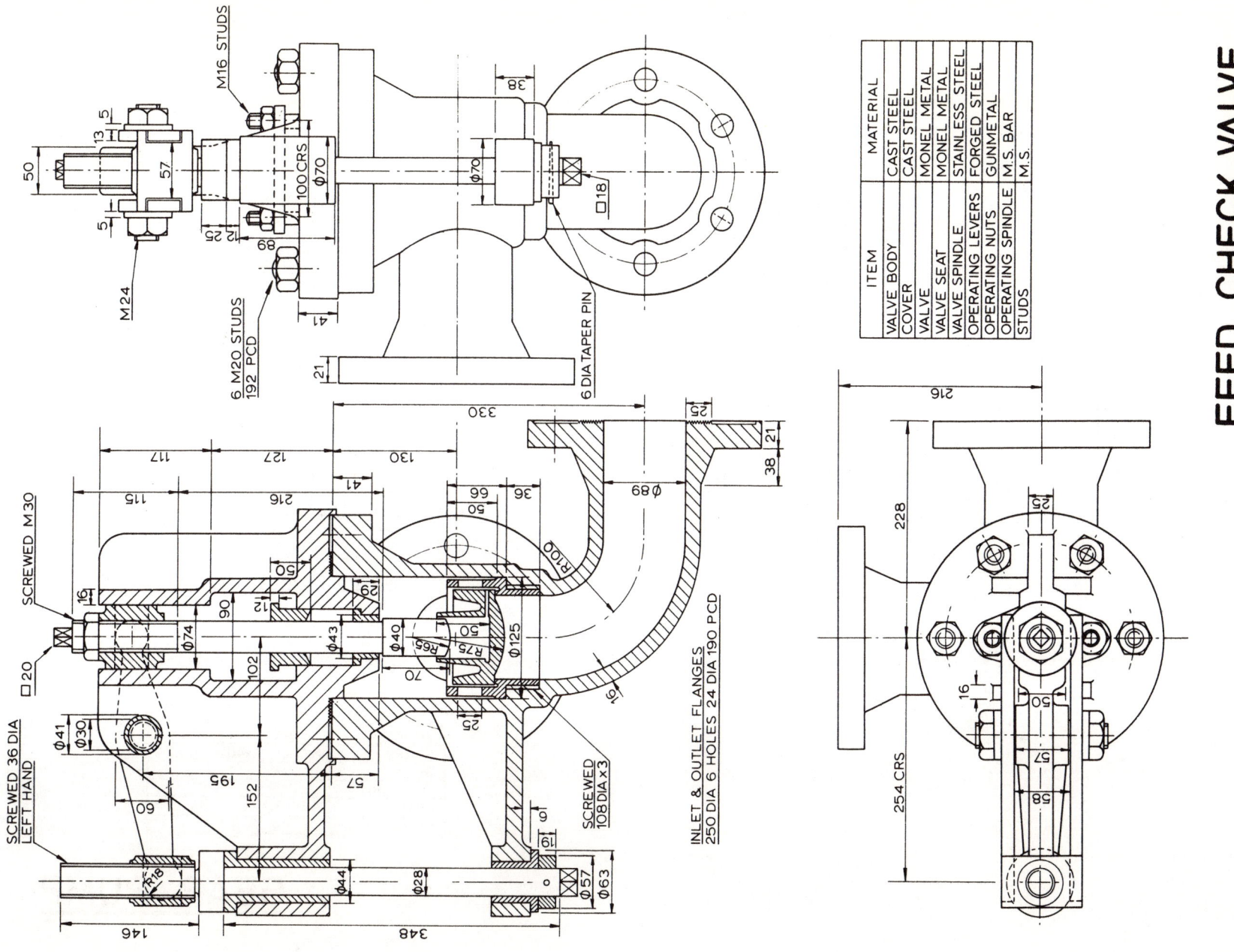

ITEM	MATERIAL
VALVE BODY	CAST STEEL
COVER	CAST STEEL
VALVE	MONEL METAL
VALVE SEAT	MONEL METAL
VALVE SPINDLE	STAINLESS STEEL
OPERATING LEVERS	FORGED STEEL
OPERATING NUTS	GUNMETAL
OPERATING SPINDLE	M.S. BAR
STUDS	M.S.

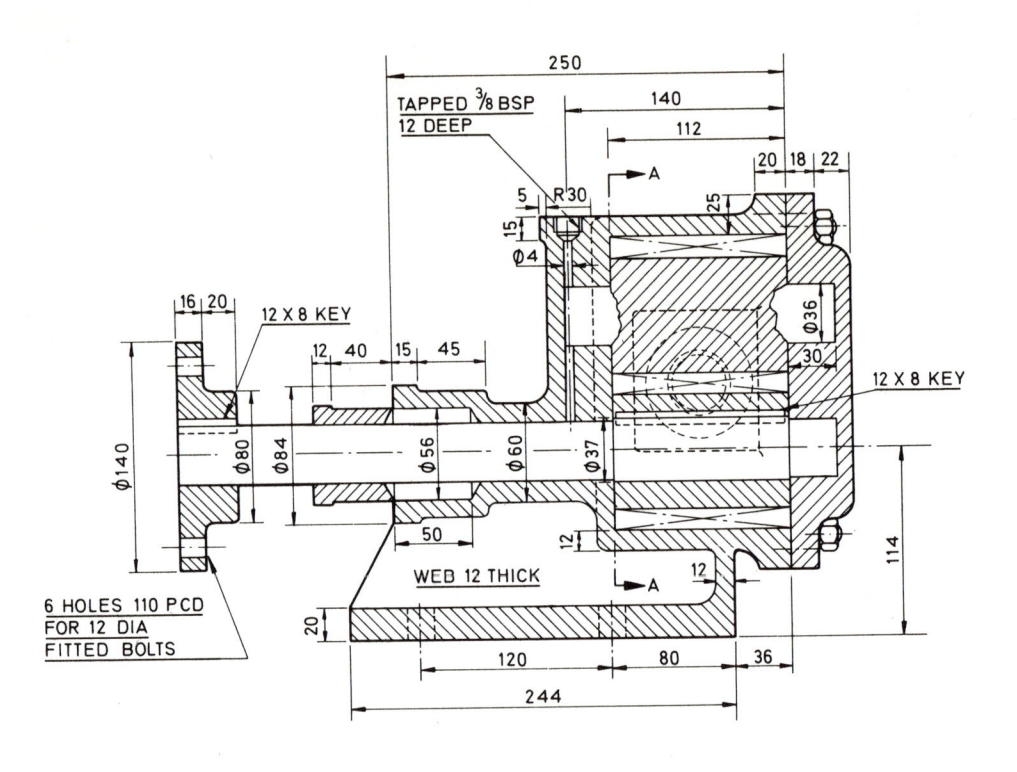

DETAILS OF INVOLUTE GEARING
PRESSURE ANGLE 20°
No. OF TEETH 14
PCD 84
ADDENDUM = MODULE = 6
DEDENDUM 7·5

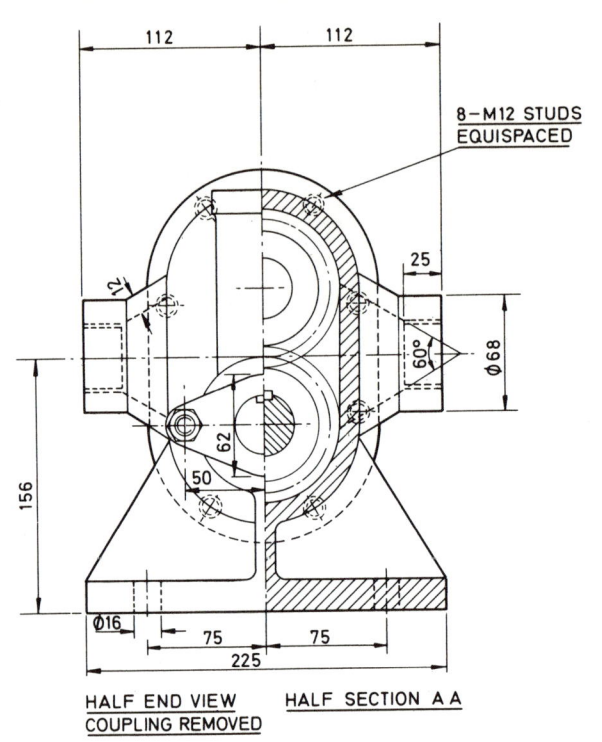

HALF END VIEW COUPLING REMOVED   HALF SECTION A A

ITEM	MATERIAL
PUMP BODY	C I
END COVER	C I
SHAFT	M S
GEARS	NICKEL CHROME STEEL
GLAND	G M
COUPLING	M S
KEYS	KEY STEEL
STUDS	M S

GEAR PUMP

STARTING AIR PILOT VALVE

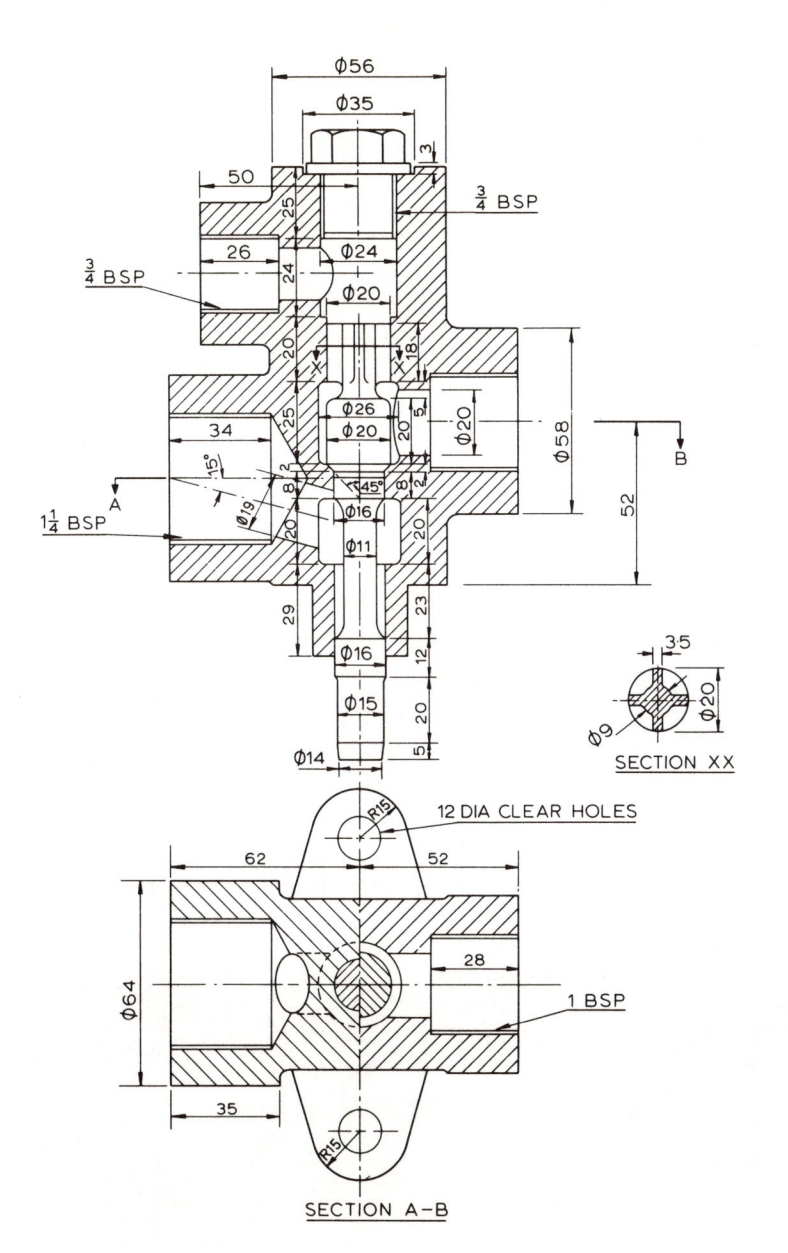

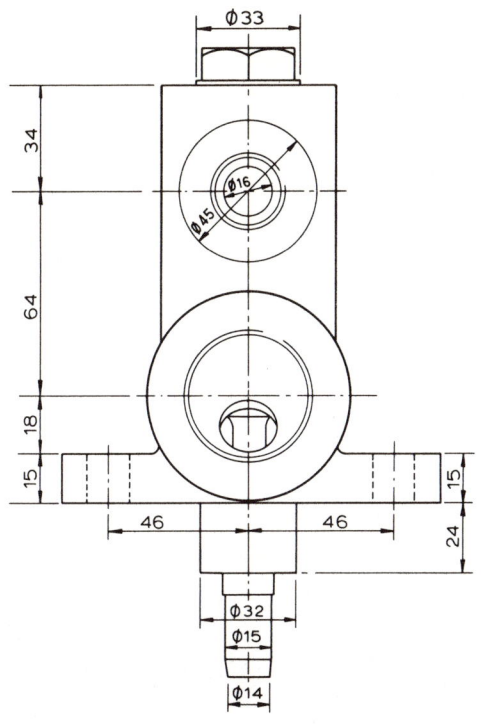

ITEM	MATERIAL
VALVE BODY	G M
VALVE	STAINLESS STEEL
PLUG	BRASS

125

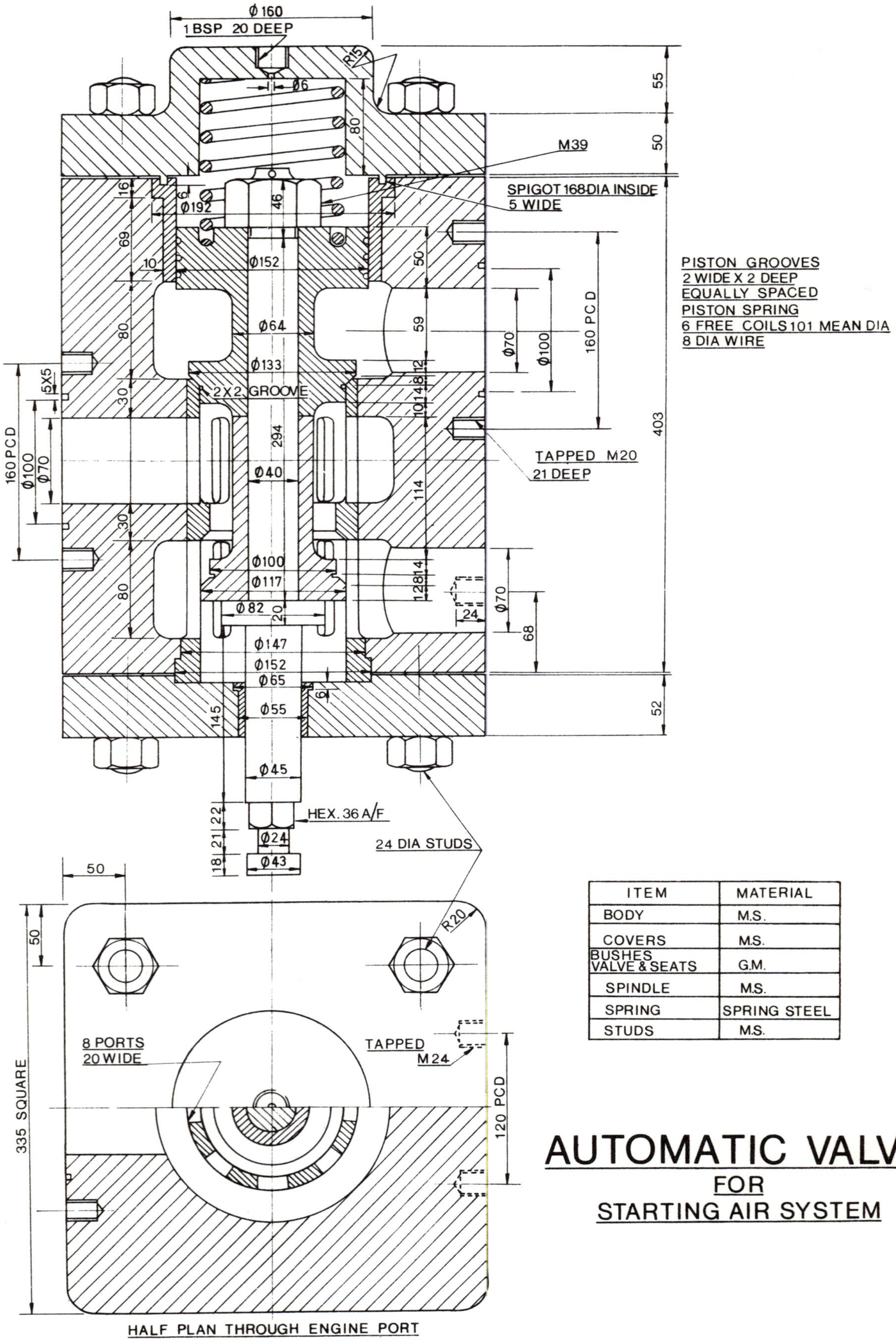

ITEM	MATERIAL
BODY	M.S.
COVERS	M.S.
BUSHES VALVE & SEATS	G.M.
SPINDLE	M.S.
SPRING	SPRING STEEL
STUDS	M.S.

AUTOMATIC VALVE
FOR
STARTING AIR SYSTEM

STARTING AIR VALVE

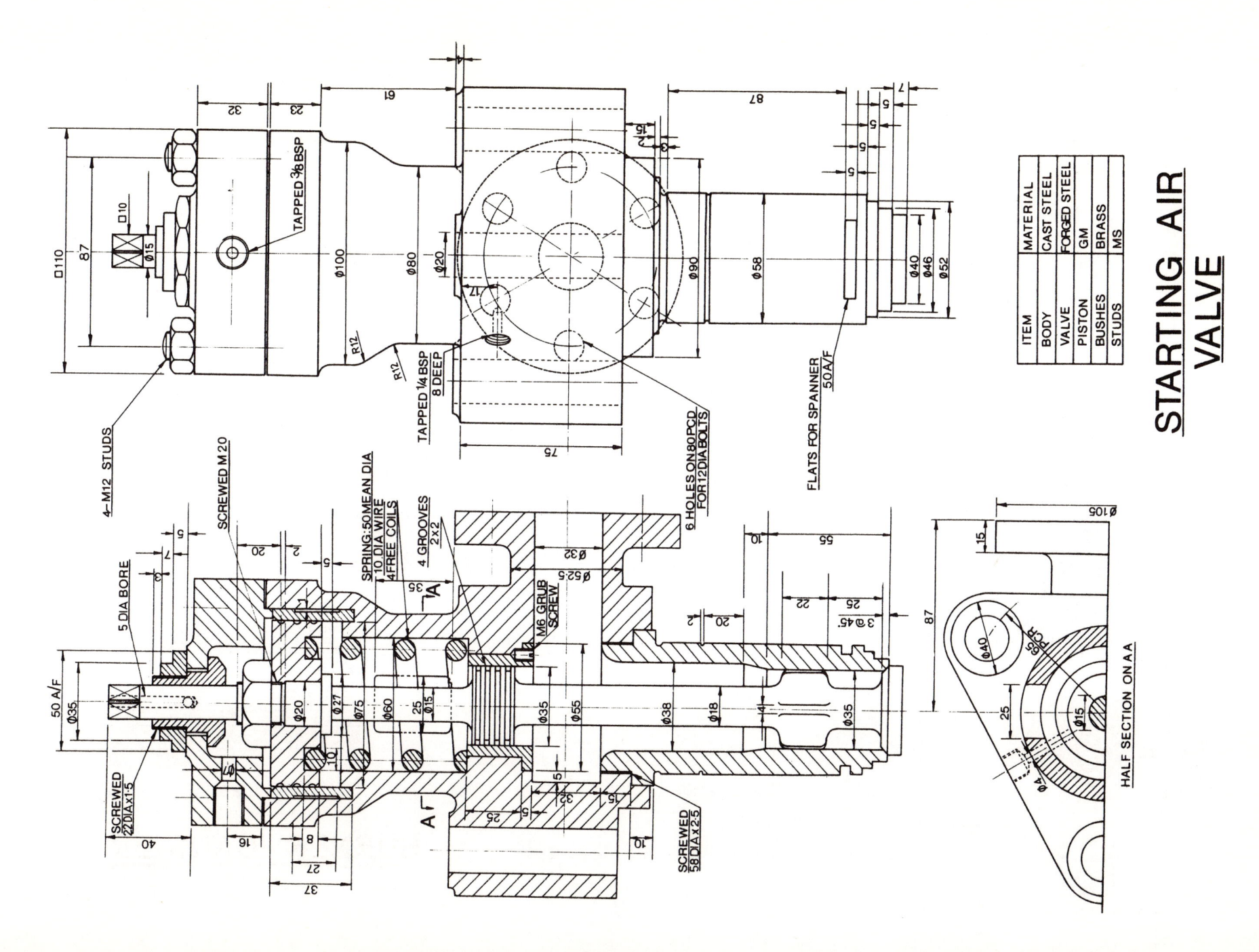

ITEM	MATERIAL
BODY	CAST STEEL
VALVE	FORGED STEEL
PISTON	GM
BUSHES	BRASS
STUDS	MS

BURNER CARRIER

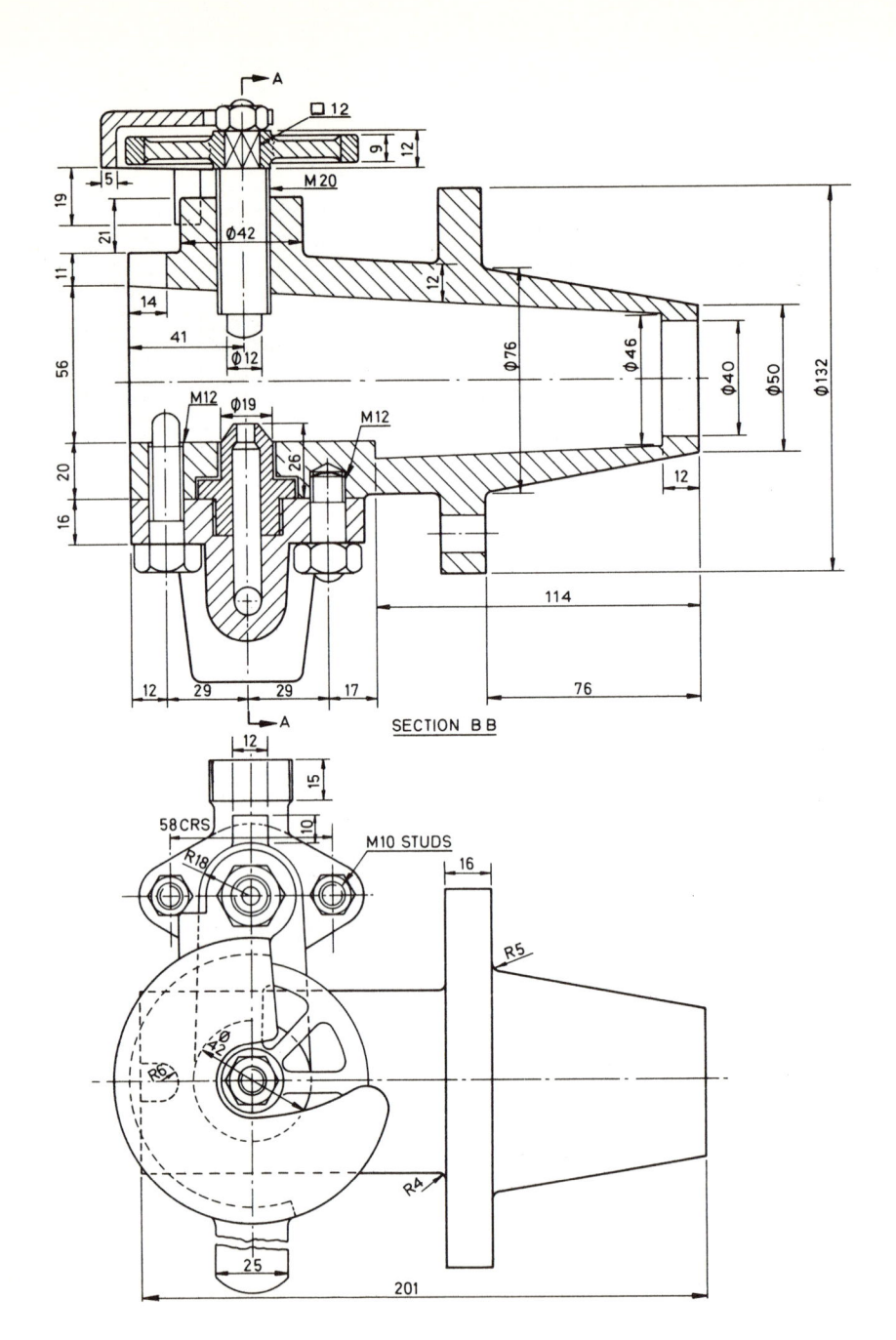

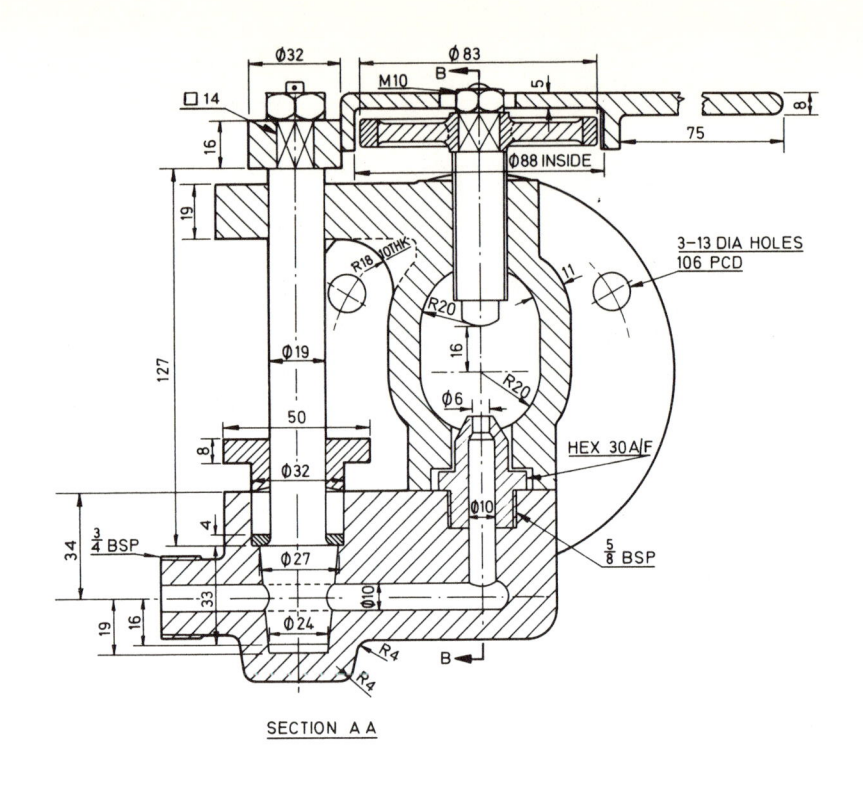

ITEM	MATERIAL
BURNER CARRIER	CAST IRON
BODY	BRONZE
COCK	M.S.
NIPPLE	M.S.
JACK SCREW	M.S.
HANDWHEEL	CAST IRON
GLAND	G.M.
SET SCREW AND STUDS	M.S.

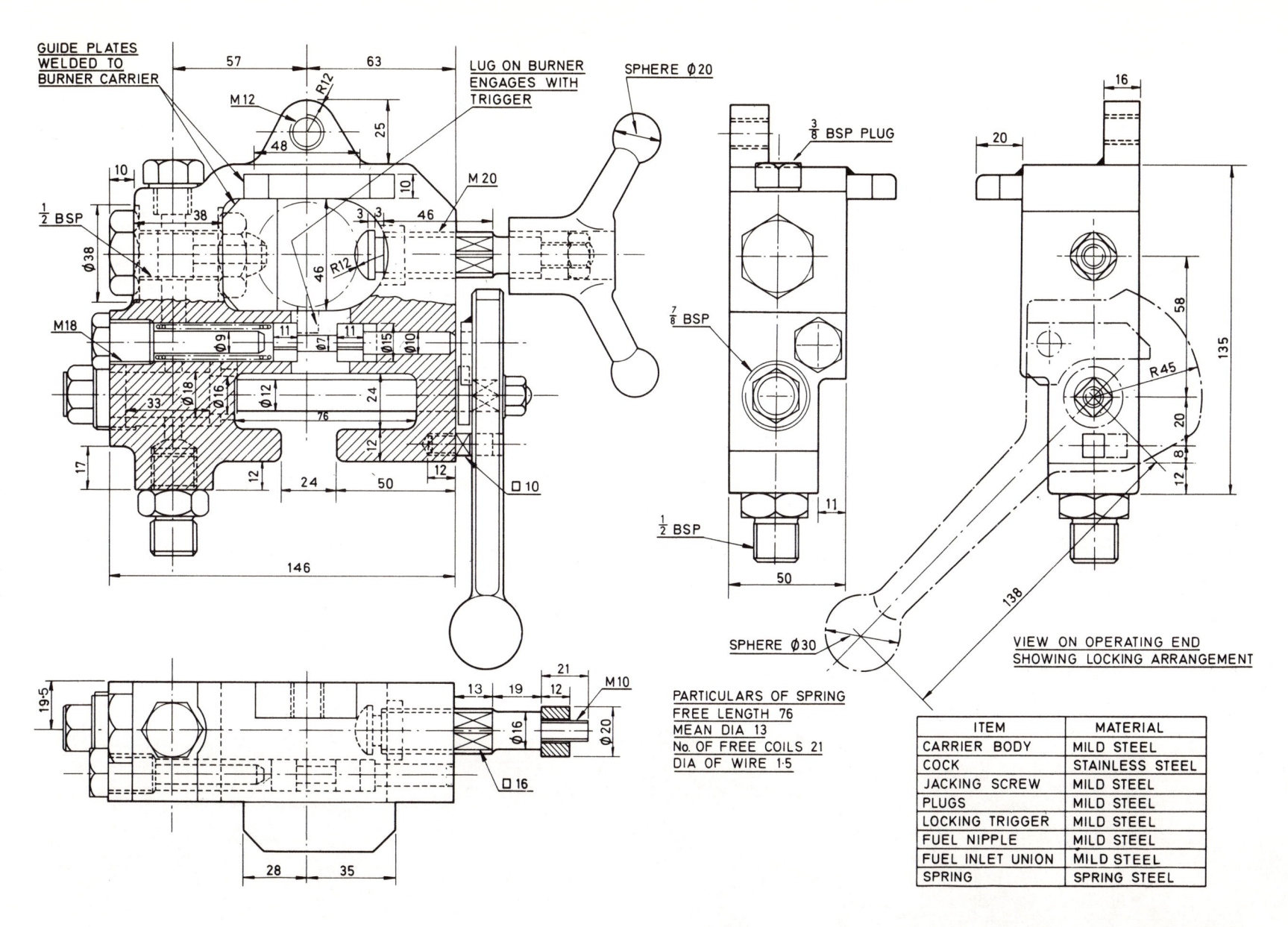

OIL FUEL BURNER CARRIER

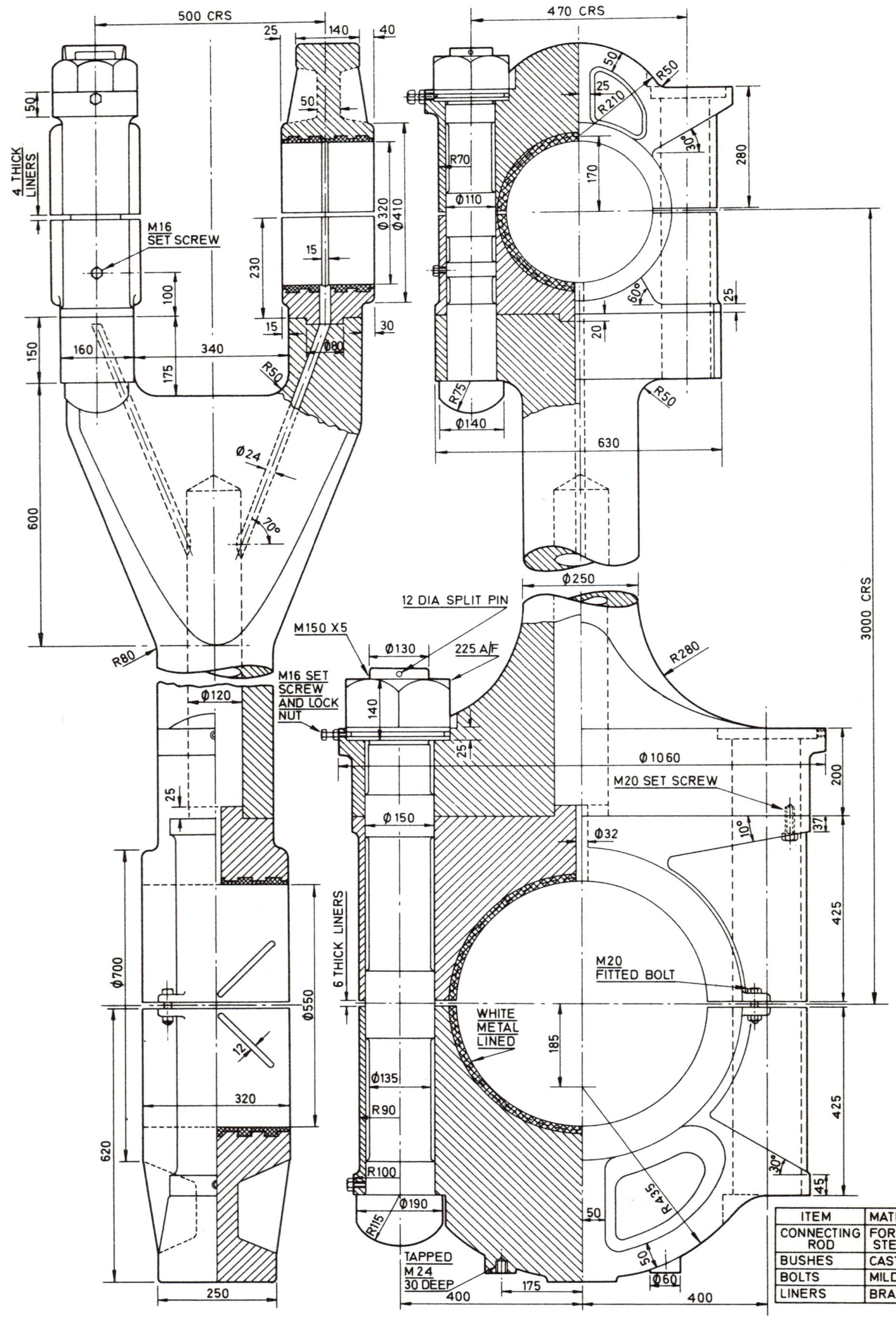

ITEM	MATERIAL
CONNECTING ROD	FORGED STEEL
BUSHES	CAST STEEL
BOLTS	MILD STEEL
LINERS	BRASS

CONNECTING ROD AND BEARINGS

GRAVITY DAVIT CENTRIFUGAL BRAKE

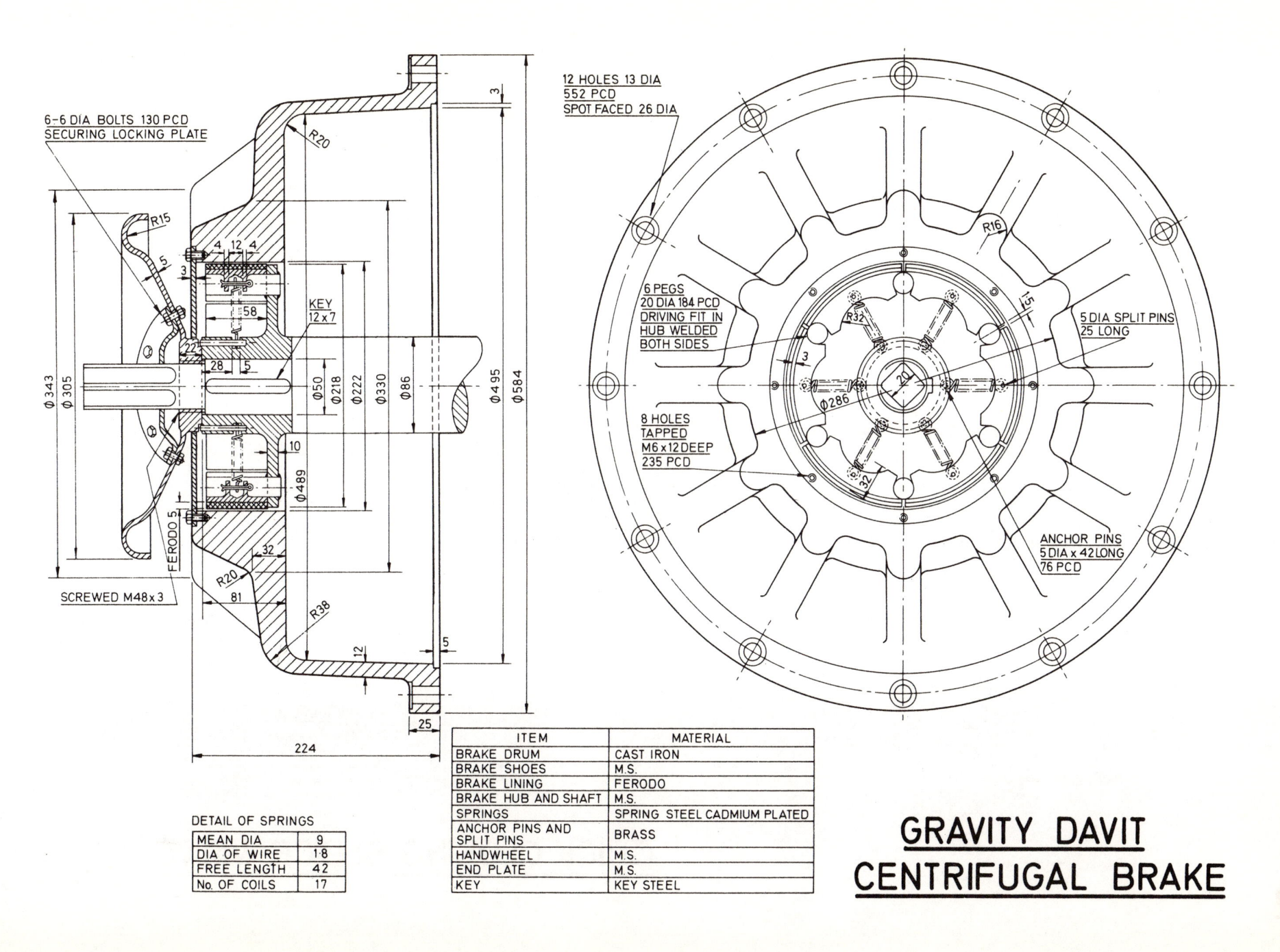

DETAIL OF SPRINGS

MEAN DIA	9
DIA OF WIRE	1.8
FREE LENGTH	42
No. OF COILS	17

ITEM	MATERIAL
BRAKE DRUM	CAST IRON
BRAKE SHOES	M.S.
BRAKE LINING	FERODO
BRAKE HUB AND SHAFT	M.S.
SPRINGS	SPRING STEEL CADMIUM PLATED
ANCHOR PINS AND SPLIT PINS	BRASS
HANDWHEEL	M.S.
END PLATE	M.S.
KEY	KEY STEEL

QUICK CLOSING SLUICE VALVE

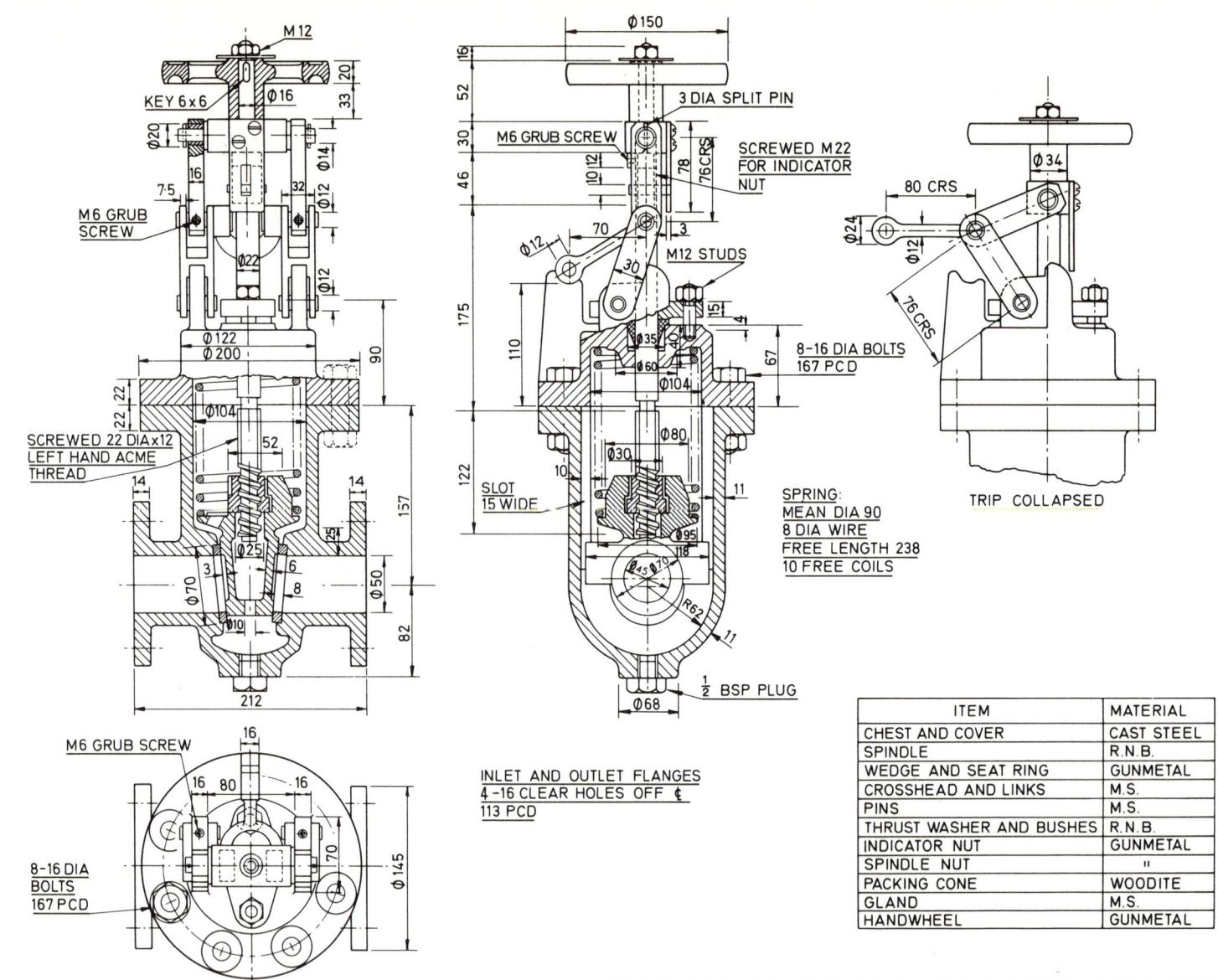

RUDDER CARRIER BEARING

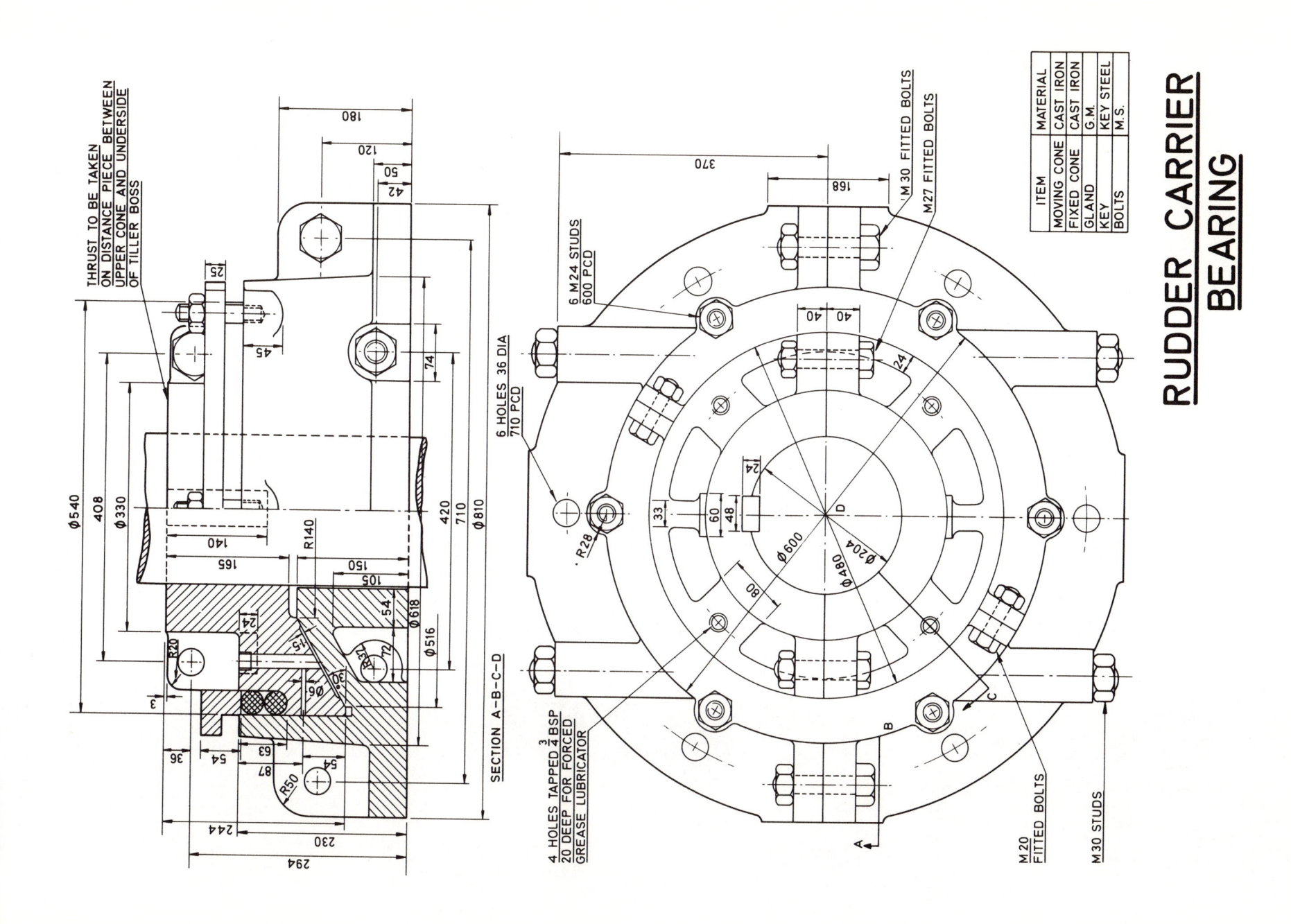

ITEM	MATERIAL
MOVING CONE	CAST IRON
FIXED CONE	CAST IRON
GLAND	G.M.
KEY	KEY STEEL
BOLTS	M.S.

THRUST TO BE TAKEN ON DISTANCE PIECE BETWEEN UPPER CONE AND UNDERSIDE OF TILLER BOSS

SECTION A-B-C-D

6 HOLES 36 DIA 710 PCD

4 HOLES TAPPED ¾ BSP 20 DEEP FOR FORCED GREASE LUBRICATOR

6 M24 STUDS 600 PCD

M27 FITTED BOLTS

M30 FITTED BOLTS

M20 FITTED BOLTS

M30 STUDS

133

REDUCING VALVE

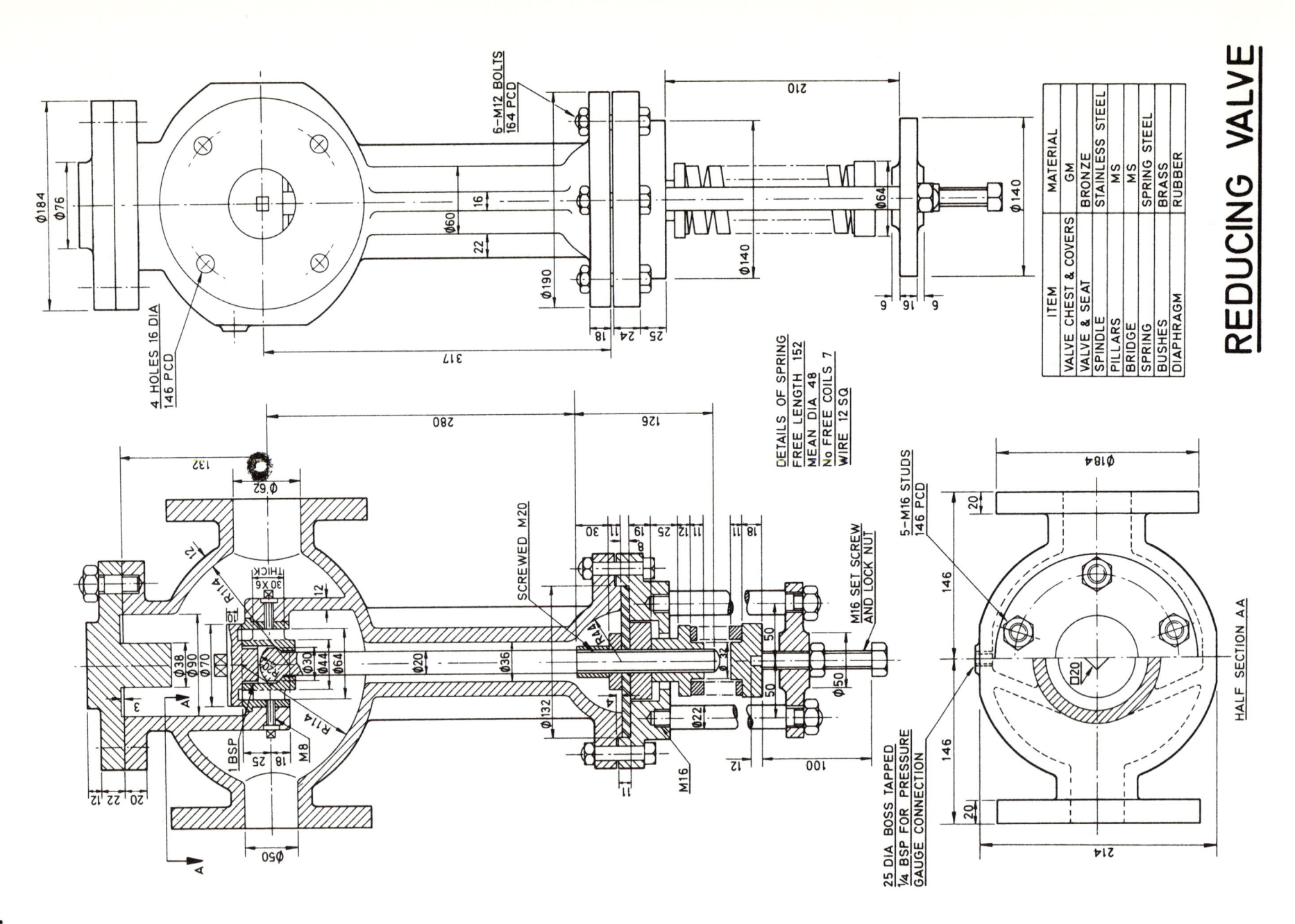

DETAILS OF SPRING
FREE LENGTH 152
MEAN DIA 48
No FREE COILS 7
WIRE 12 SQ

25 DIA BOSS TAPPED
¼ BSP FOR PRESSURE
GAUGE CONNECTION

HALF SECTION AA

ITEM	MATERIAL
VALVE CHEST & COVERS	GM
VALVE & SEAT	BRONZE
SPINDLE	STAINLESS STEEL
PILLARS	MS
BRIDGE	MS
SPRING	SPRING STEEL
BUSHES	BRASS
DIAPHRAGM	RUBBER

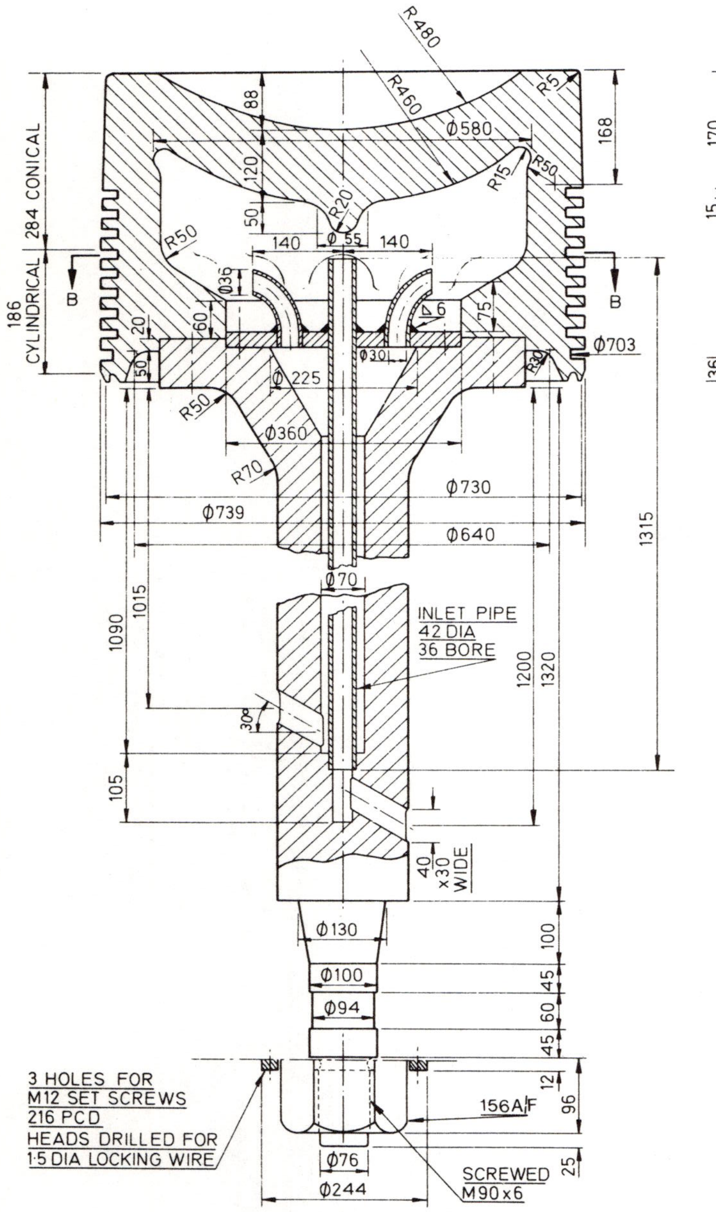

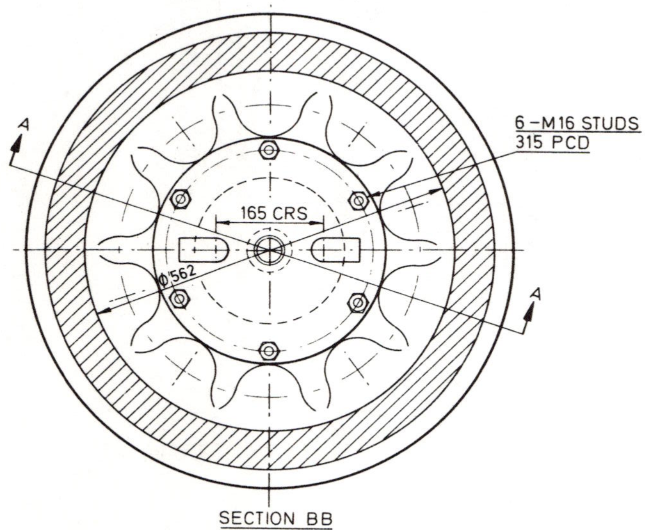

SECTION BB

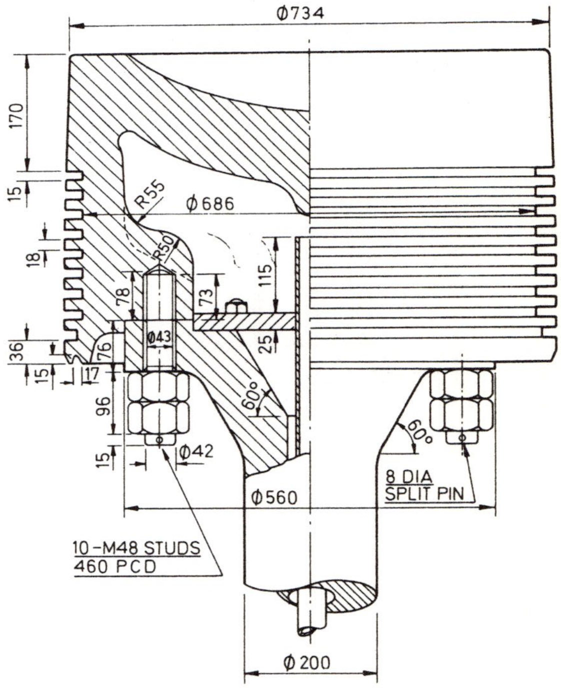

HALF SECTION AA

ITEM	MATERIAL
PISTON	CAST IRON
PISTON ROD	FORGED STEEL
COOLING ASSEMBLY	M S
PISTON ROD NUT	WROUGHT IRON
LOCKING PLATE	M S
STUDS AND NUTS	M S
SPLIT PINS	M S

4-STROKE PISTON AND ROD

UPPER PISTON AND ROD

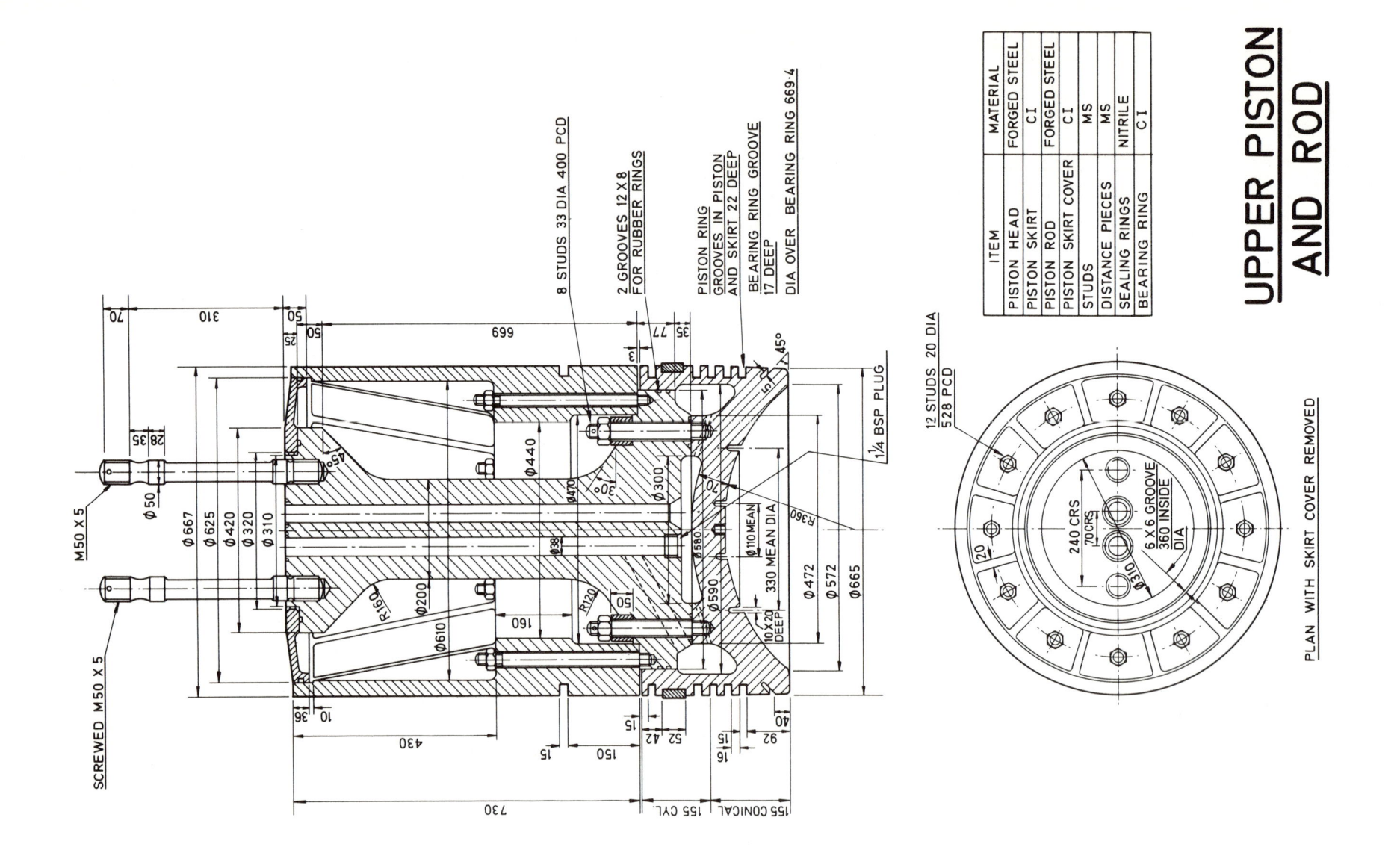

ITEM	MATERIAL
PISTON HEAD	FORGED STEEL
PISTON SKIRT	CI
PISTON ROD	FORGED STEEL
PISTON SKIRT COVER	CI
STUDS	MS
DISTANCE PIECES	MS
SEALING RINGS	NITRILE
BEARING RING	CI

PLAN WITH SKIRT COVER REMOVED

TELEMOTOR RECEIVER

Details of Springs

MEAN DIA	50
DIA OF WIRE	10
No. OF FREE COILS	35
FREE LENGTH	536

Materials

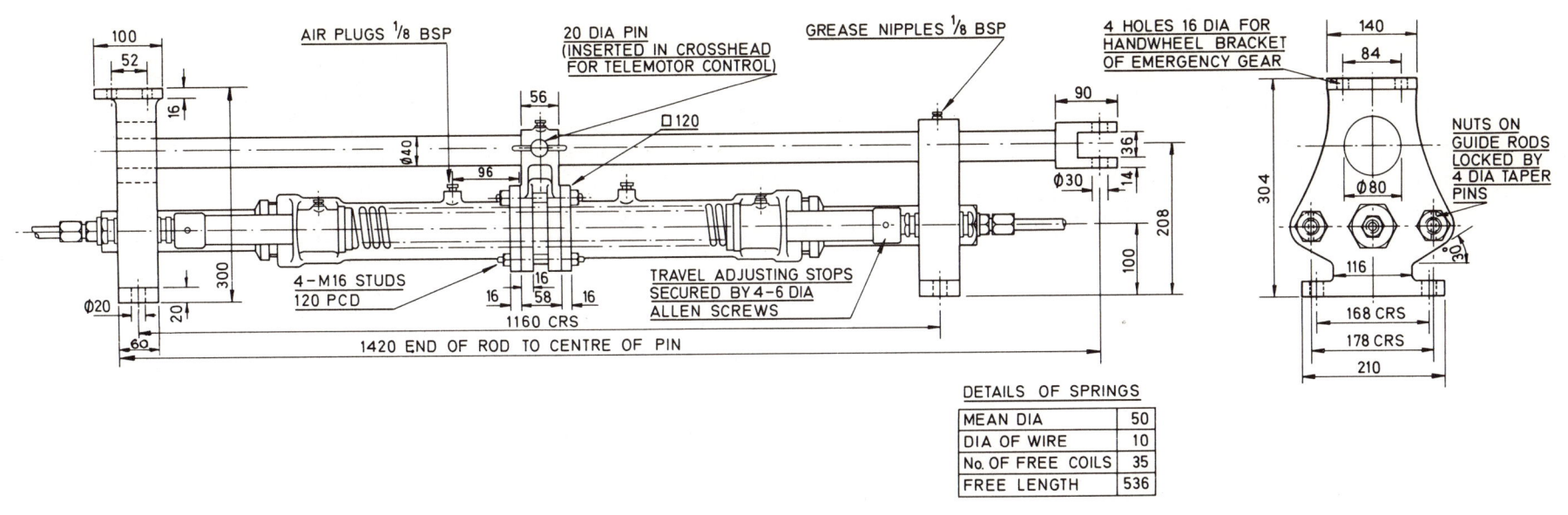

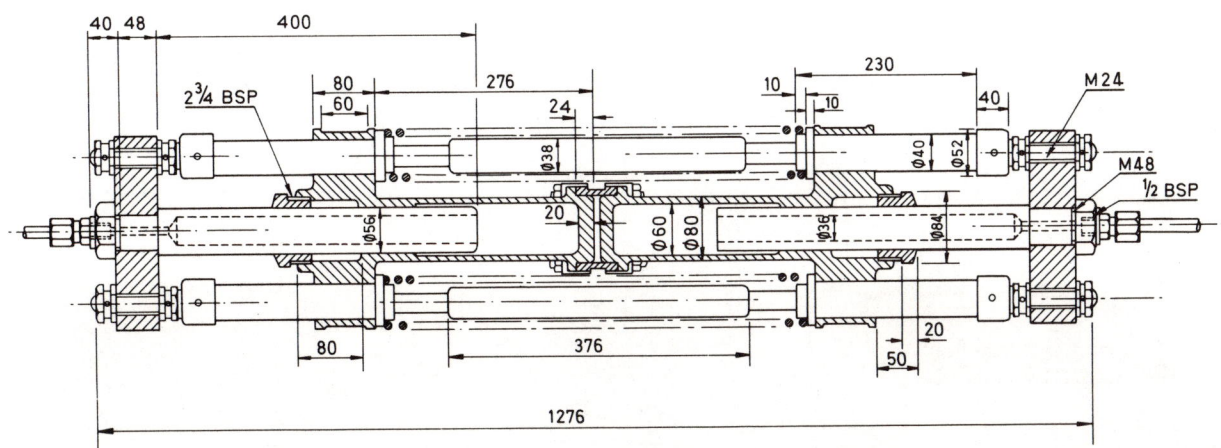

ITEM	MATERIAL
CYLINDERS	CAST BRONZE
RAMS	MS CHROM. PLATED
PUMP OPERATING ROD	MILD STEEL
RAM BRACKETS	C I
CROSSHEAD	BRASS
GUIDE RODS	M S
GLAND NUTS	BRONZE
SPRINGS	SPRING STEEL
SPRING SUPPORTS	STEEL TUBE

VALVE ACTUATOR

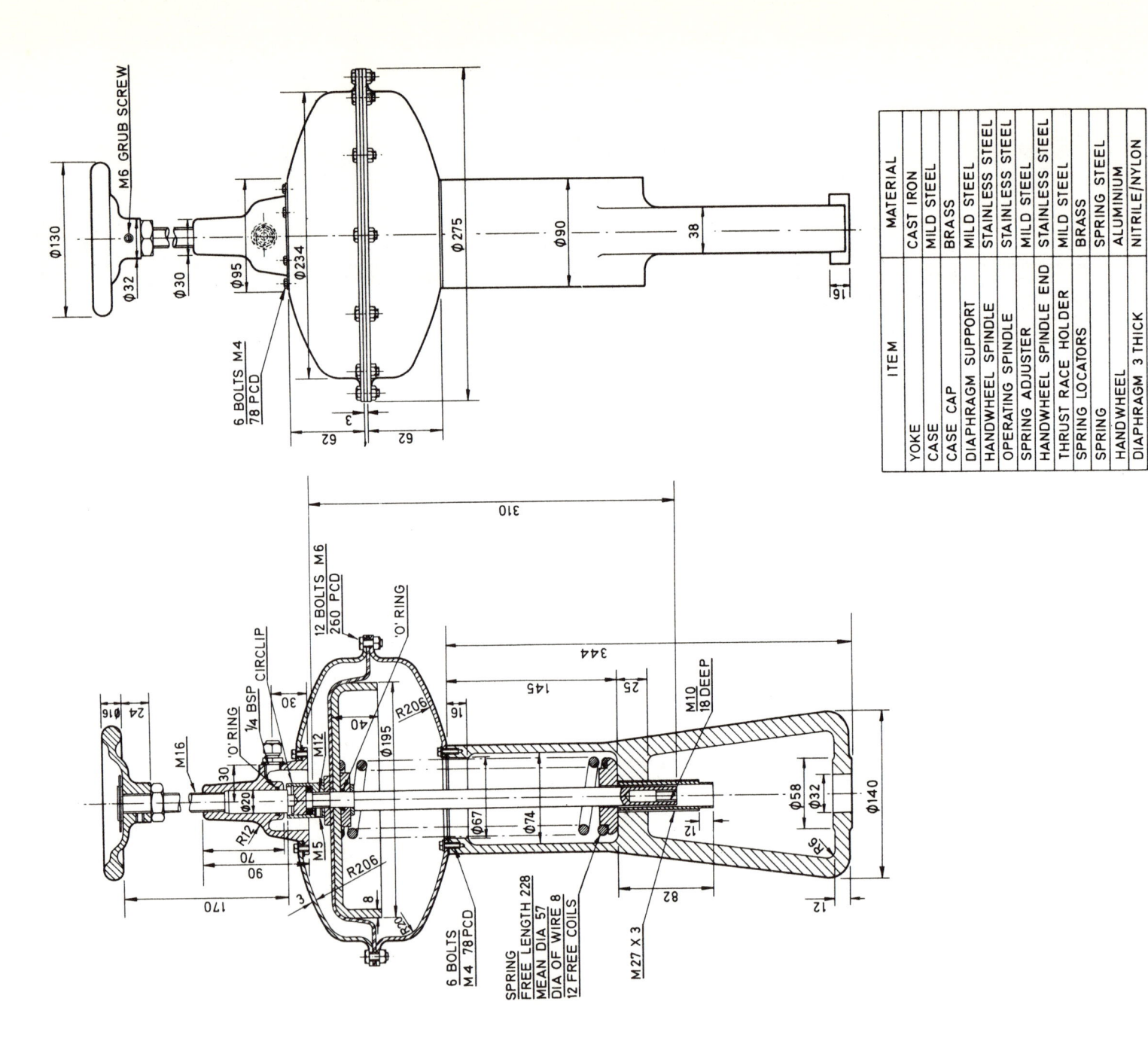

ITEM	MATERIAL
YOKE	CAST IRON
CASE	MILD STEEL
CASE CAP	BRASS
DIAPHRAGM SUPPORT	MILD STEEL
HANDWHEEL SPINDLE	STAINLESS STEEL
OPERATING SPINDLE	STAINLESS STEEL
SPRING ADJUSTER	MILD STEEL
HANDWHEEL SPINDLE END	STAINLESS STEEL
THRUST RACE HOLDER	MILD STEEL
SPRING LOCATORS	BRASS
SPRING	SPRING STEEL
HANDWHEEL	ALUMINIUM
DIAPHRAGM 3 THICK	NITRILE/NYLON

138

GENERATOR PEDESTAL BEARING

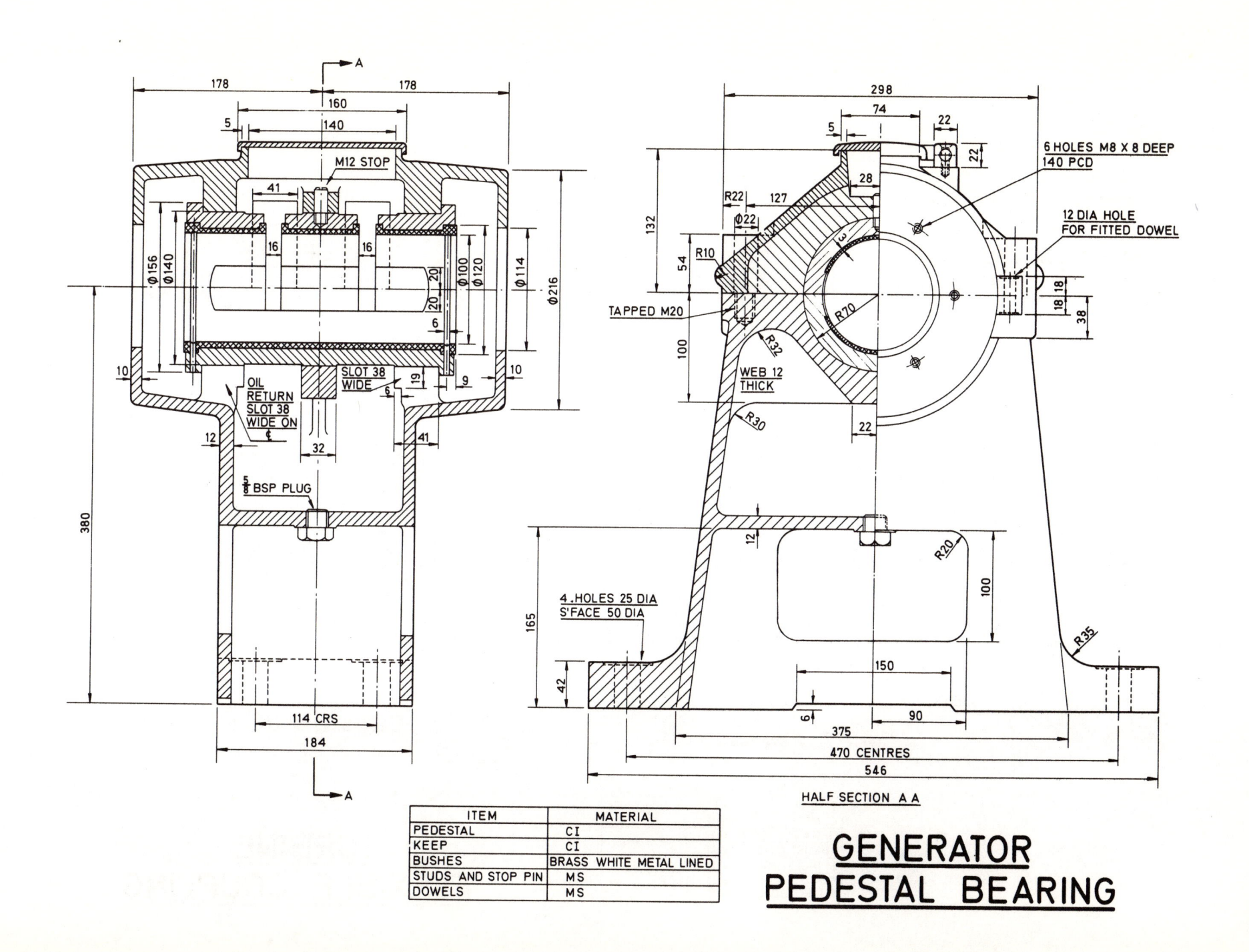

HALF SECTION A A

ITEM	MATERIAL
PEDESTAL	CI
KEEP	CI
BUSHES	BRASS WHITE METAL LINED
STUDS AND STOP PIN	MS
DOWELS	MS

TURBINE FLEXIBLE COUPLING

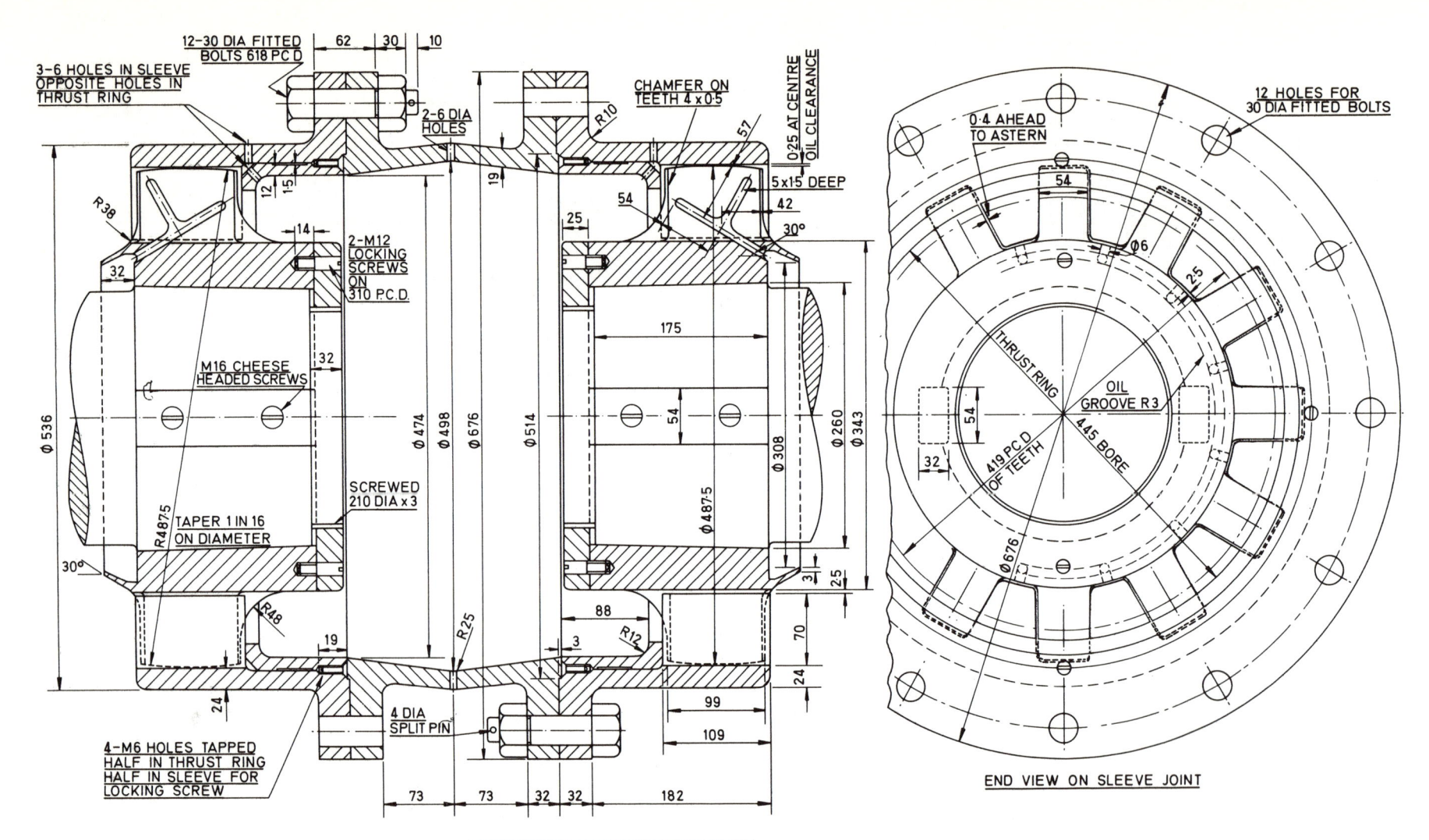

ITEM	MATERIAL
SLEEVE	MED. CARBON STEEL
CLAW	M.S.
DISTANCE PIECE	M.S.
THRUST RING	G.M.
SPINDLE NUT	WROUGHT IRON
KEYS	KEY STEEL
FITTED BOLTS	M.S.

TURBINE FLEXIBLE COUPLING

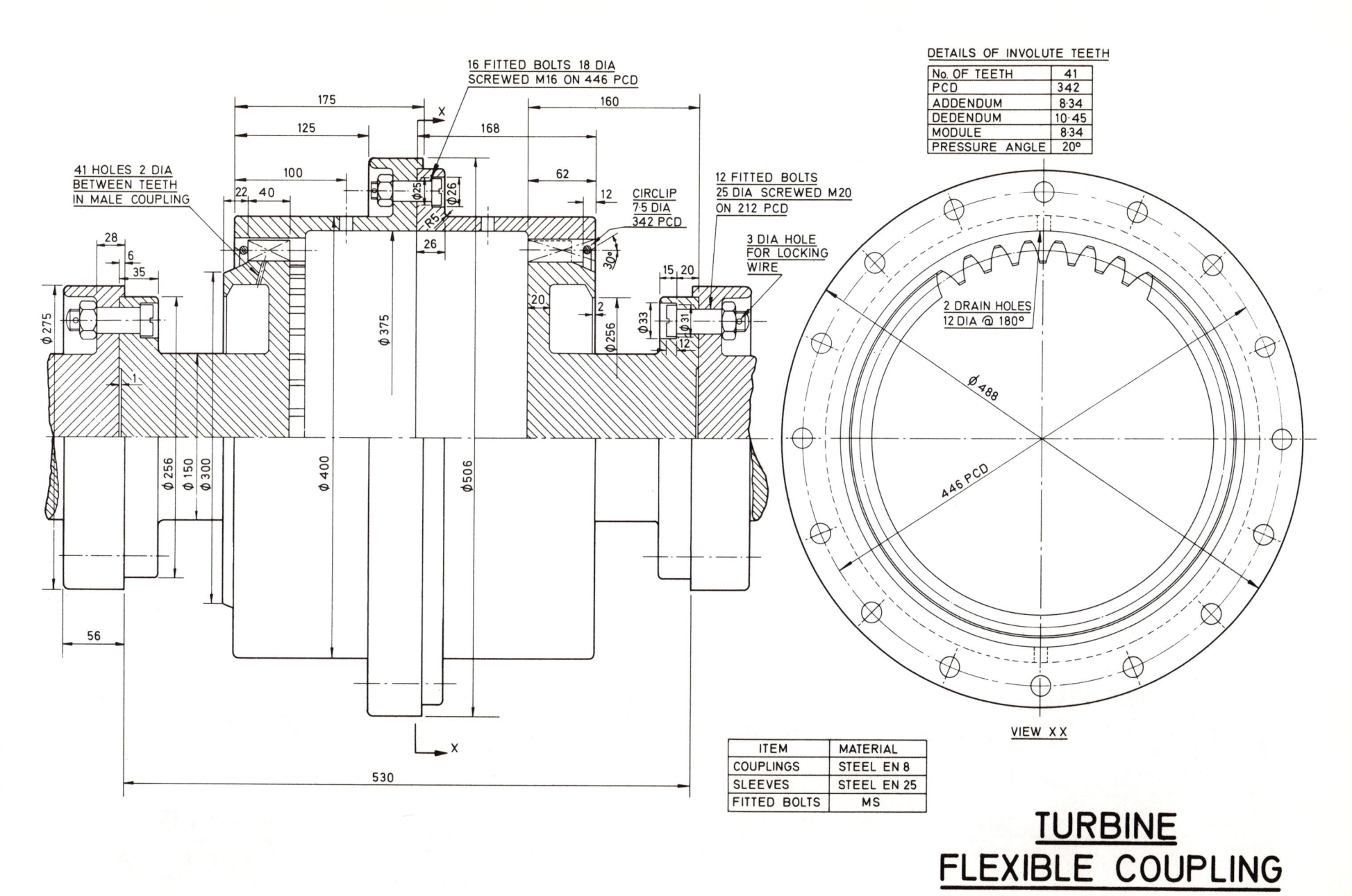

DETAILS OF INVOLUTE TEETH

No. OF TEETH	41
PCD	342
ADDENDUM	8·34
DEDENDUM	10·45
MODULE	8·34
PRESSURE ANGLE	20°

ITEM	MATERIAL
COUPLINGS	STEEL EN 8
SLEEVES	STEEL EN 25
FITTED BOLTS	MS

16 FITTED BOLTS 18 DIA SCREWED M16 ON 446 PCD

12 FITTED BOLTS 25 DIA SCREWED M20 ON 212 PCD

41 HOLES 2 DIA BETWEEN TEETH IN MALE COUPLING

CIRCLIP 7·5 DIA 342 PCD

3 DIA HOLE FOR LOCKING WIRE

2 DRAIN HOLES 12 DIA @ 180°

VIEW XX

141

HIGH LIFT SAFETY VALVE

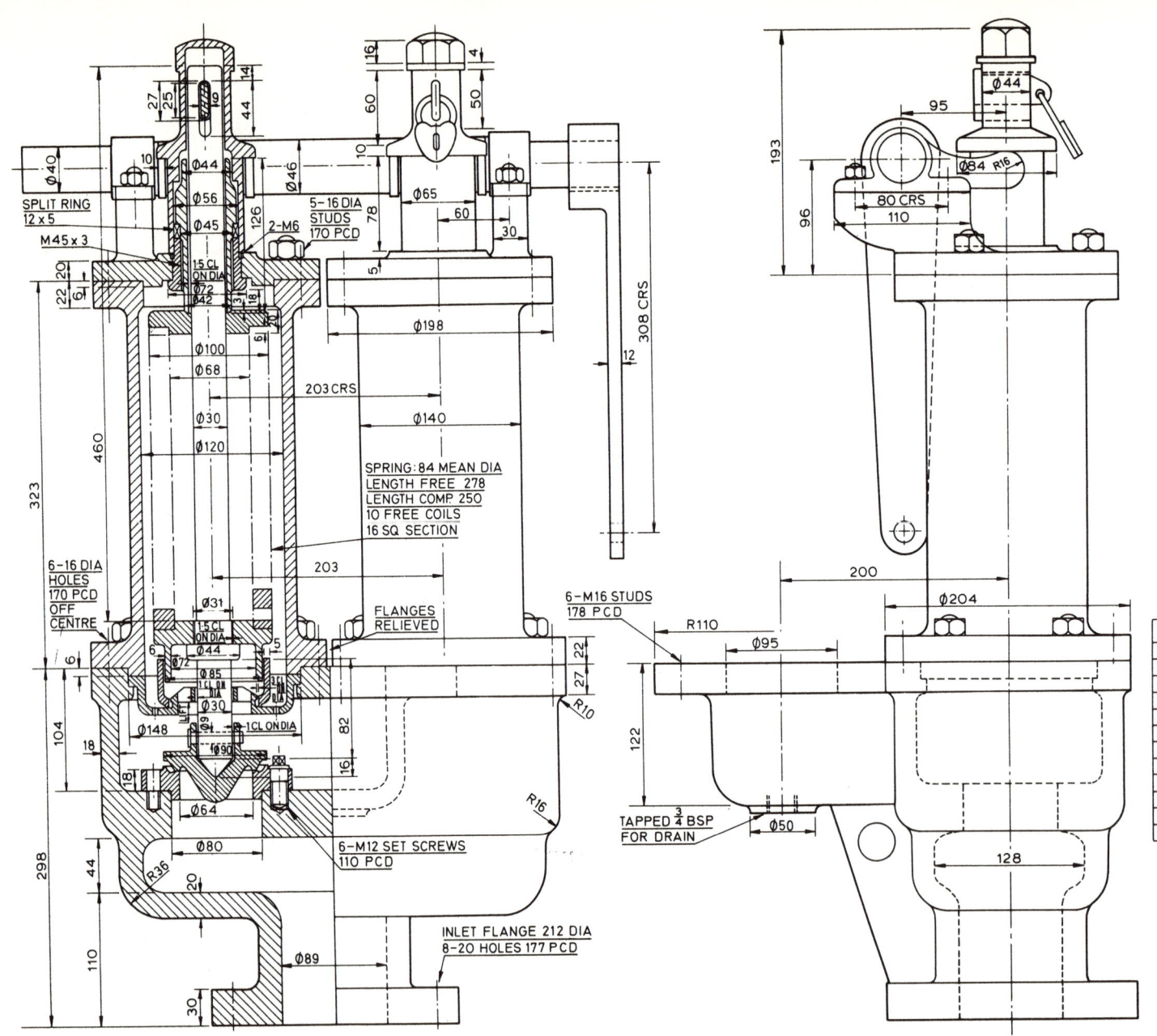

ITEM	MATERIAL
CHEST	CAST STEEL
CASING	CAST IRON
COVER	CAST IRON
VALVE AND SEAT	ALUMINIUM BRONZE
DIAPHRAGM	GUN METAL
LOOSE RING	GUN METAL
PISTON	GUN METAL
SPINDLE	E.N.B.
SPRING	SPRING STEEL
COVER BUSH	BRASS
COMPRESSION RING	BRASS
EASING SHAFT	MILD STEEL

FULL BORE SAFETY VALVE

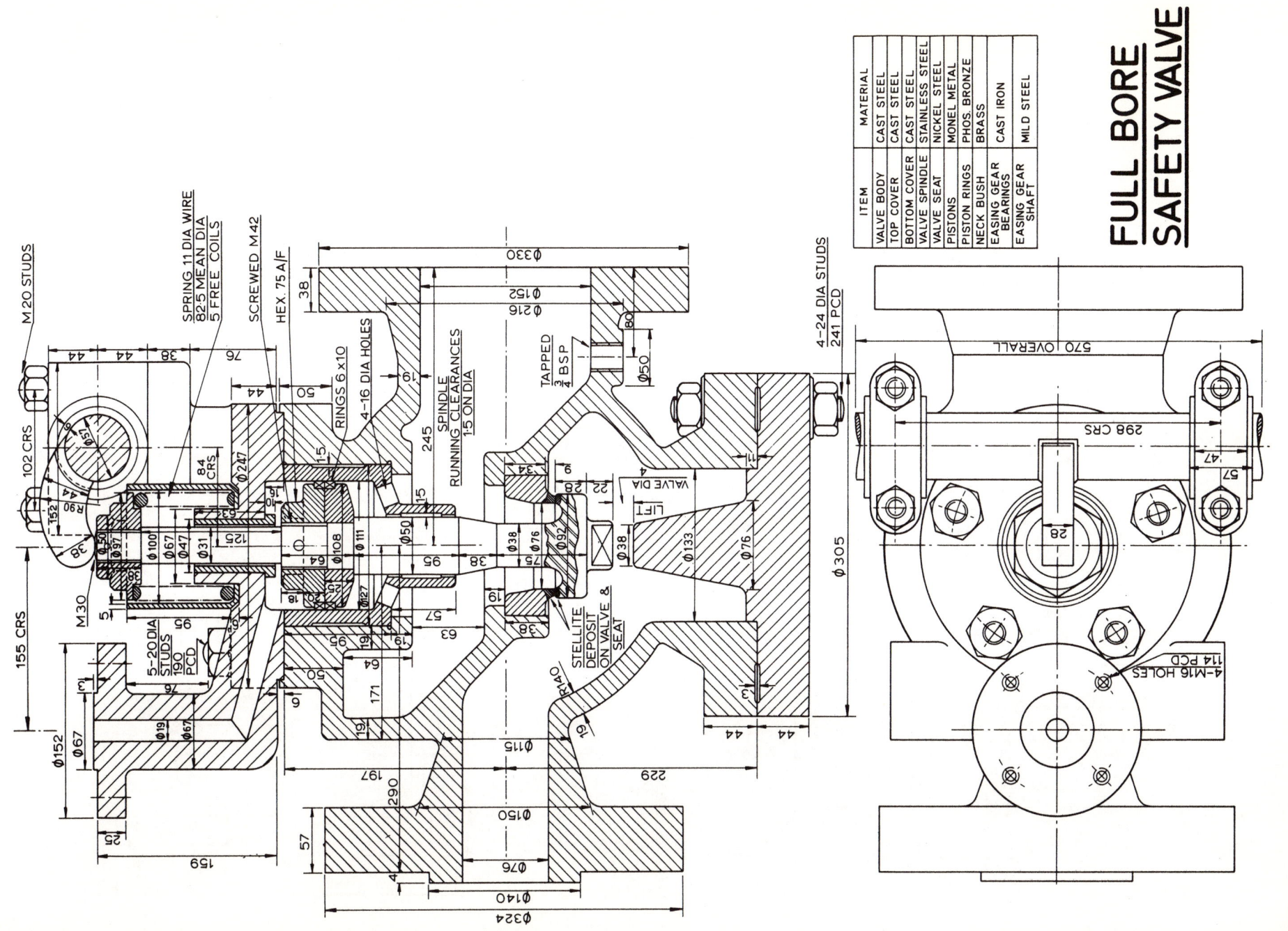

ITEM	MATERIAL
VALVE BODY	CAST STEEL
TOP COVER	CAST STEEL
BOTTOM COVER	CAST STEEL
VALVE SPINDLE	STAINLESS STEEL
VALVE SEAT	NICKEL STEEL
PISTONS	MONEL METAL
PISTON RINGS	PHOS. BRONZE
NECK BUSH	BRASS
EASING GEAR BEARINGS	CAST IRON
EASING GEAR SHAFT	MILD STEEL

FLOW REGULATOR

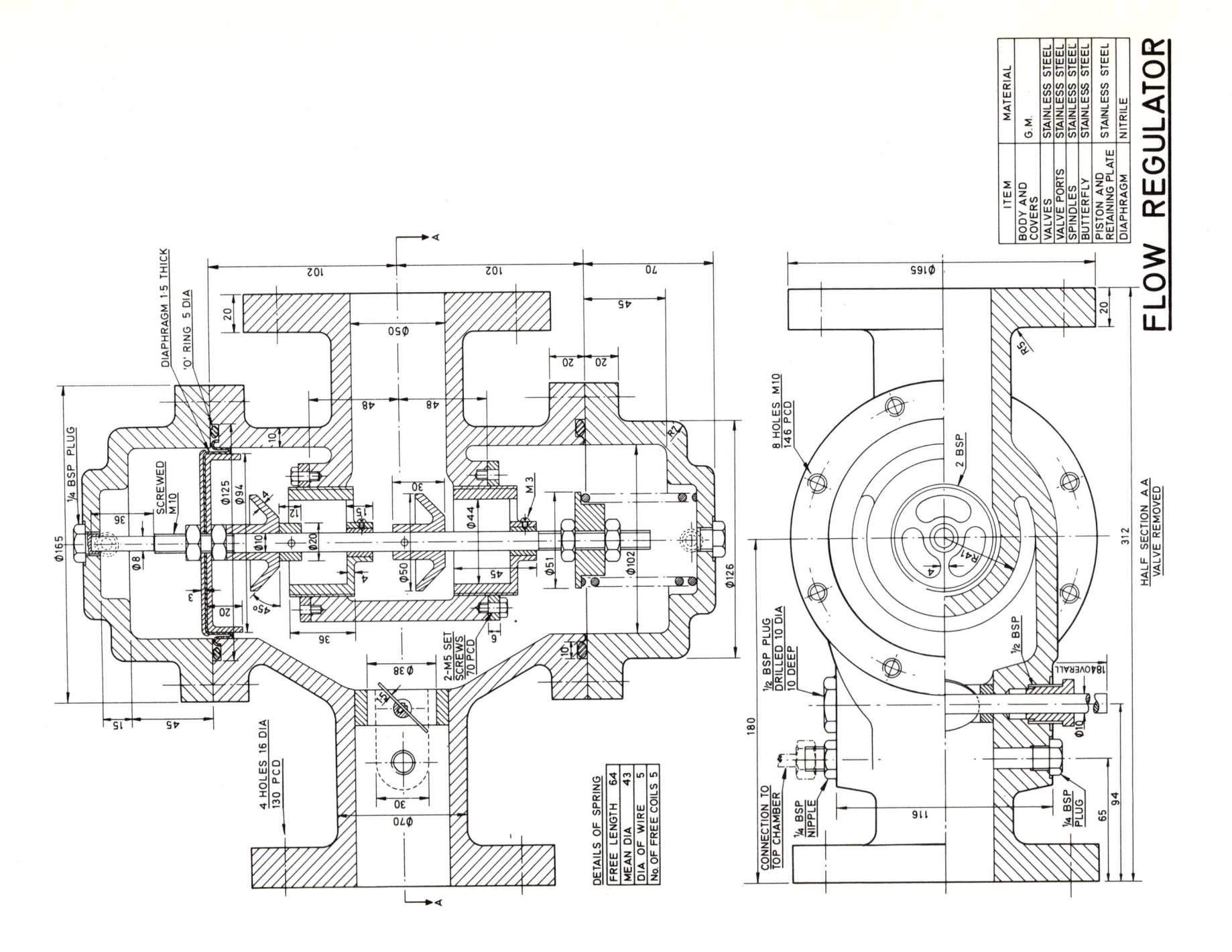

ITEM	MATERIAL
BODY AND COVERS	G.M.
VALVES	STAINLESS STEEL
VALVE PORTS	STAINLESS STEEL
SPINDLES	STAINLESS STEEL
BUTTERFLY	STAINLESS STEEL
PISTON AND RETAINING PLATE	STAINLESS STEEL
DIAPHRAGM	NITRILE

DETAILS OF SPRING

FREE LENGTH	64
MEAN DIA	43
DIA OF WIRE	5
No OF FREE COILS	5

HALF SECTION A A
VALVE REMOVED

FUEL VALVE

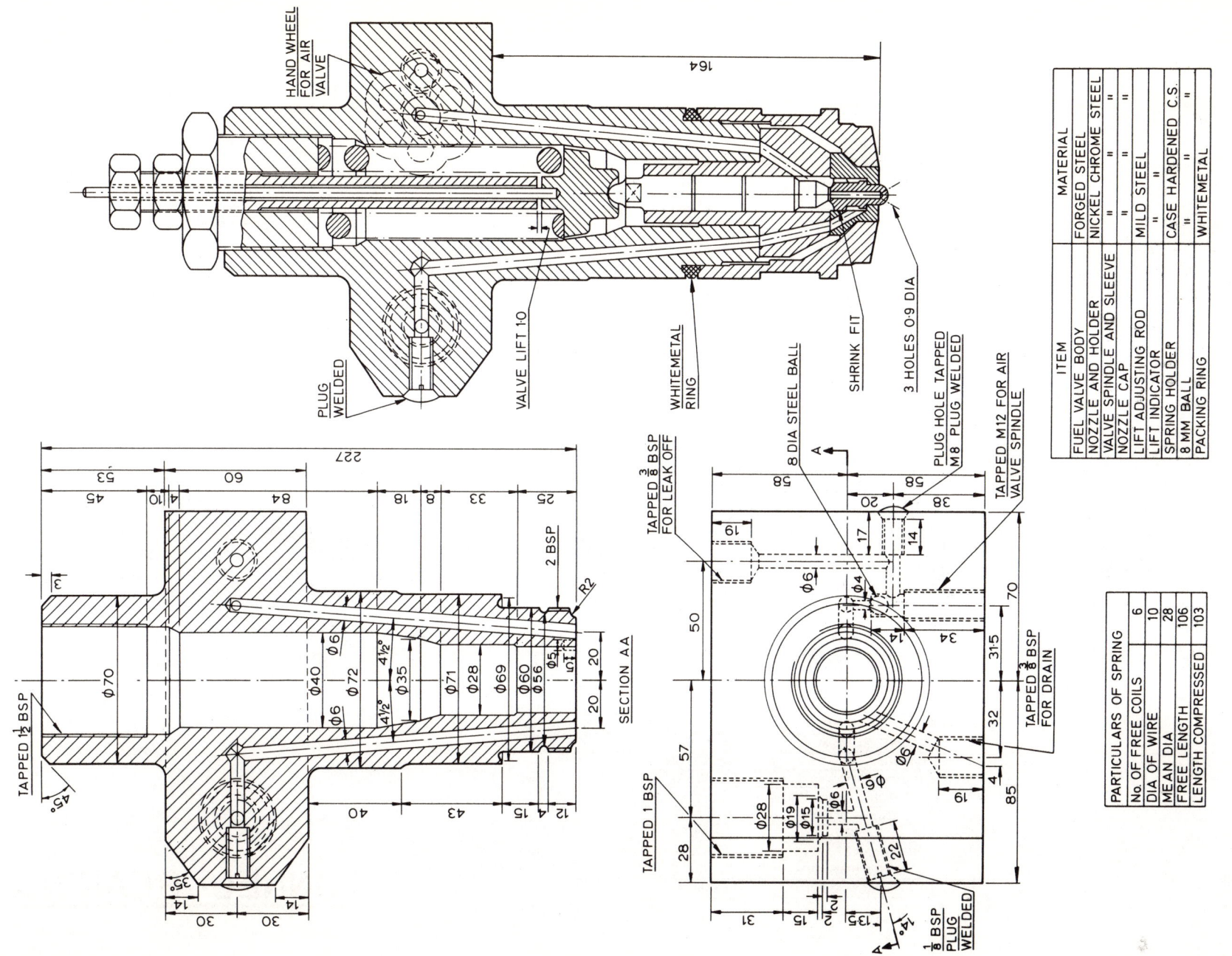

ITEM	MATERIAL
FUEL VALVE BODY	FORGED STEEL
NOZZLE AND HOLDER	NICKEL CHROME STEEL
VALVE SPINDLE AND SLEEVE	" "
NOZZLE CAP	MILD STEEL
LIFT ADJUSTING ROD	" "
LIFT INDICATOR	" "
SPRING HOLDER	CASE HARDENED C.S.
8 MM BALL	" "
PACKING RING	WHITEMETAL

PARTICULARS OF SPRING	
No. OF FREE COILS	6
DIA OF WIRE	10
MEAN DIA	28
FREE LENGTH	106
LENGTH COMPRESSED	103

145

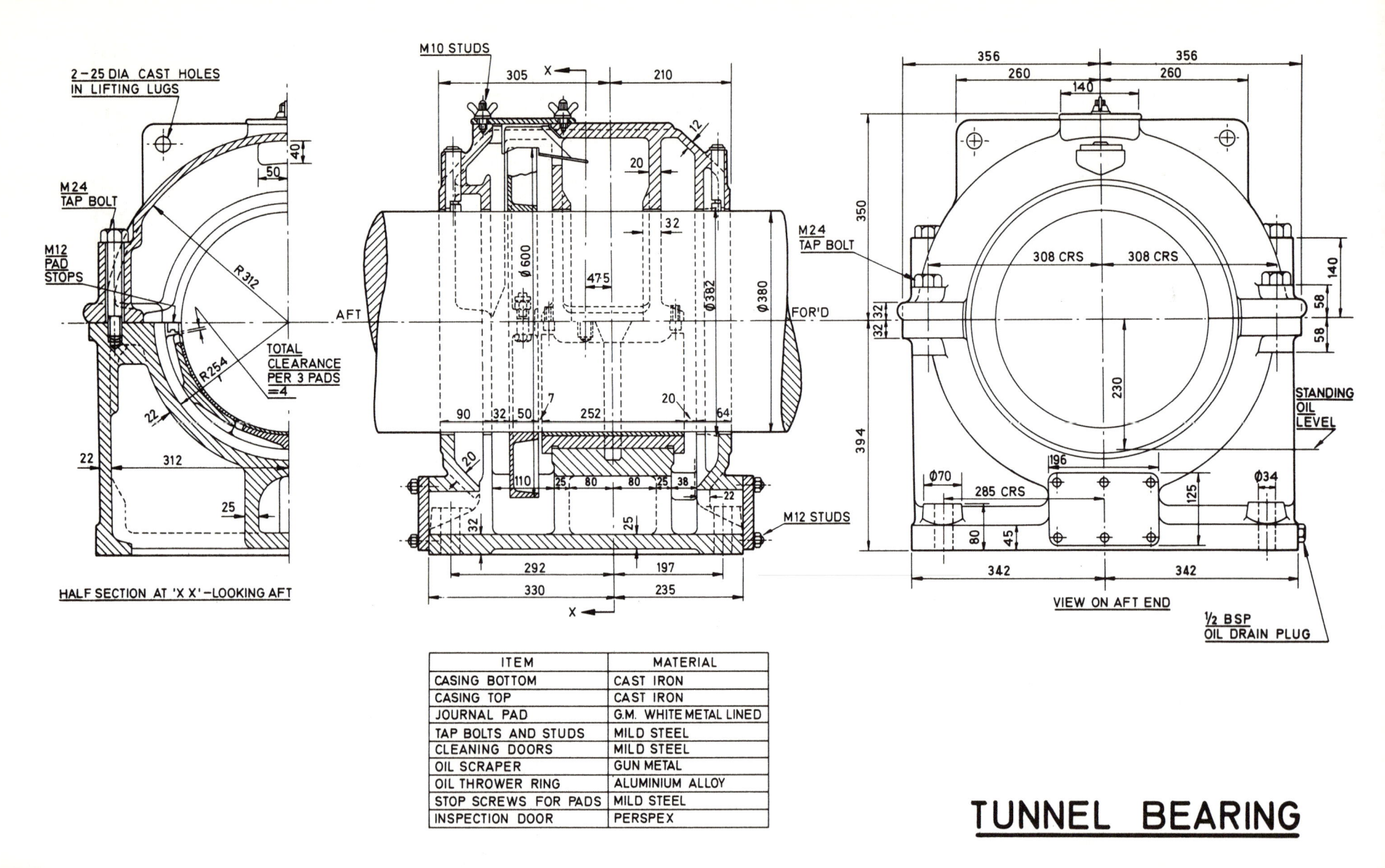

ITEM	MATERIAL
CASING BOTTOM	CAST IRON
CASING TOP	CAST IRON
JOURNAL PAD	G.M. WHITE METAL LINED
TAP BOLTS AND STUDS	MILD STEEL
CLEANING DOORS	MILD STEEL
OIL SCRAPER	GUN METAL
OIL THROWER RING	ALUMINIUM ALLOY
STOP SCREWS FOR PADS	MILD STEEL
INSPECTION DOOR	PERSPEX

TUNNEL BEARING

AIR INLET VALVE

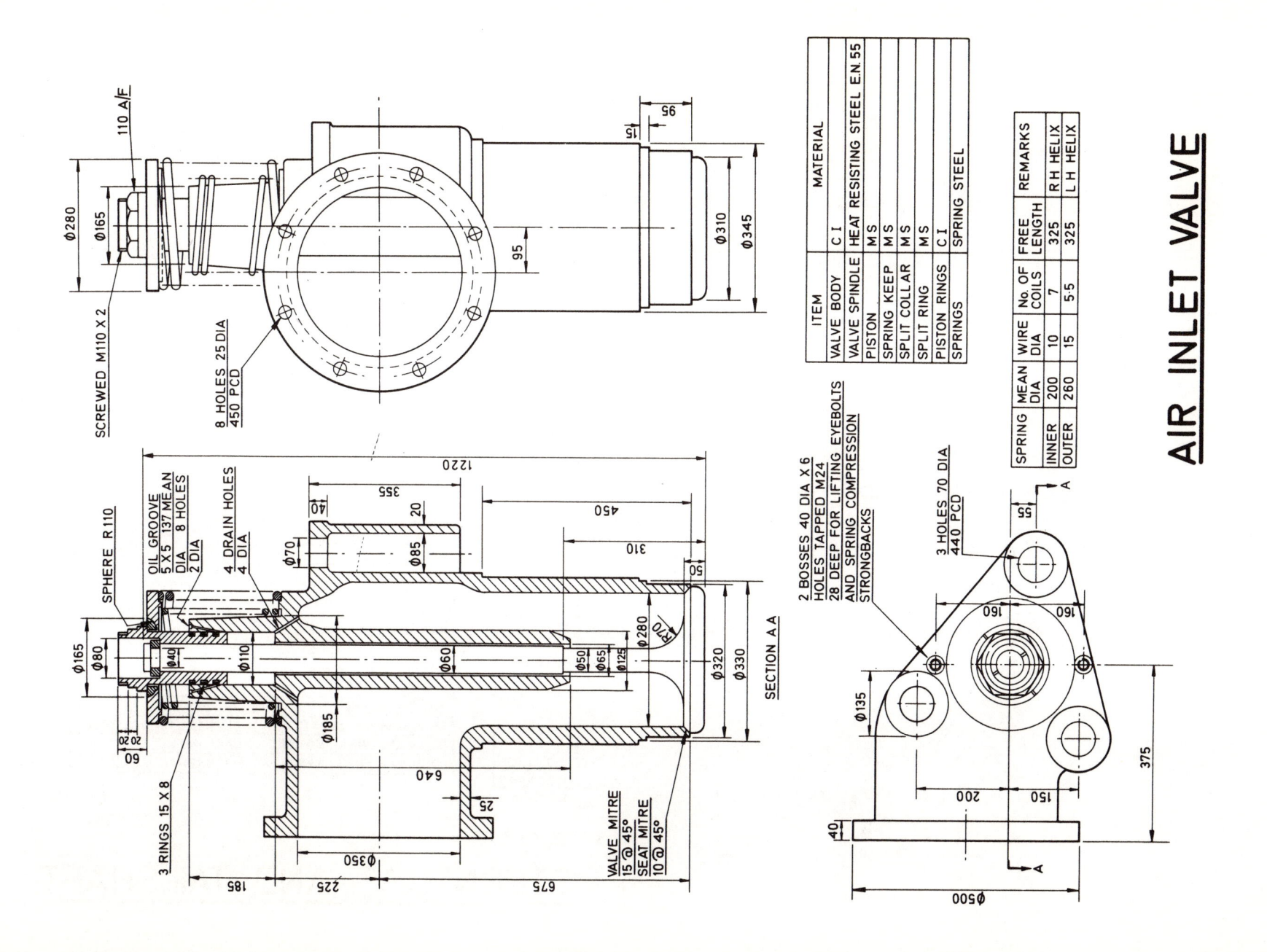

ITEM	MATERIAL
VALVE BODY	C I
VALVE SPINDLE	HEAT RESISTING STEEL E.N. 55
PISTON	M S
SPRING KEEP	M S
SPLIT COLLAR	M S
SPLIT RING	M S
PISTON RINGS	C I
SPRINGS	SPRING STEEL

SPRING	MEAN DIA	WIRE DIA	No. OF COILS	FREE LENGTH	REMARKS
INNER	200	10	7	325	R H HELIX
OUTER	260	15	5.5	325	L H HELIX

147

STERNTUBE AND TAILSHAFT

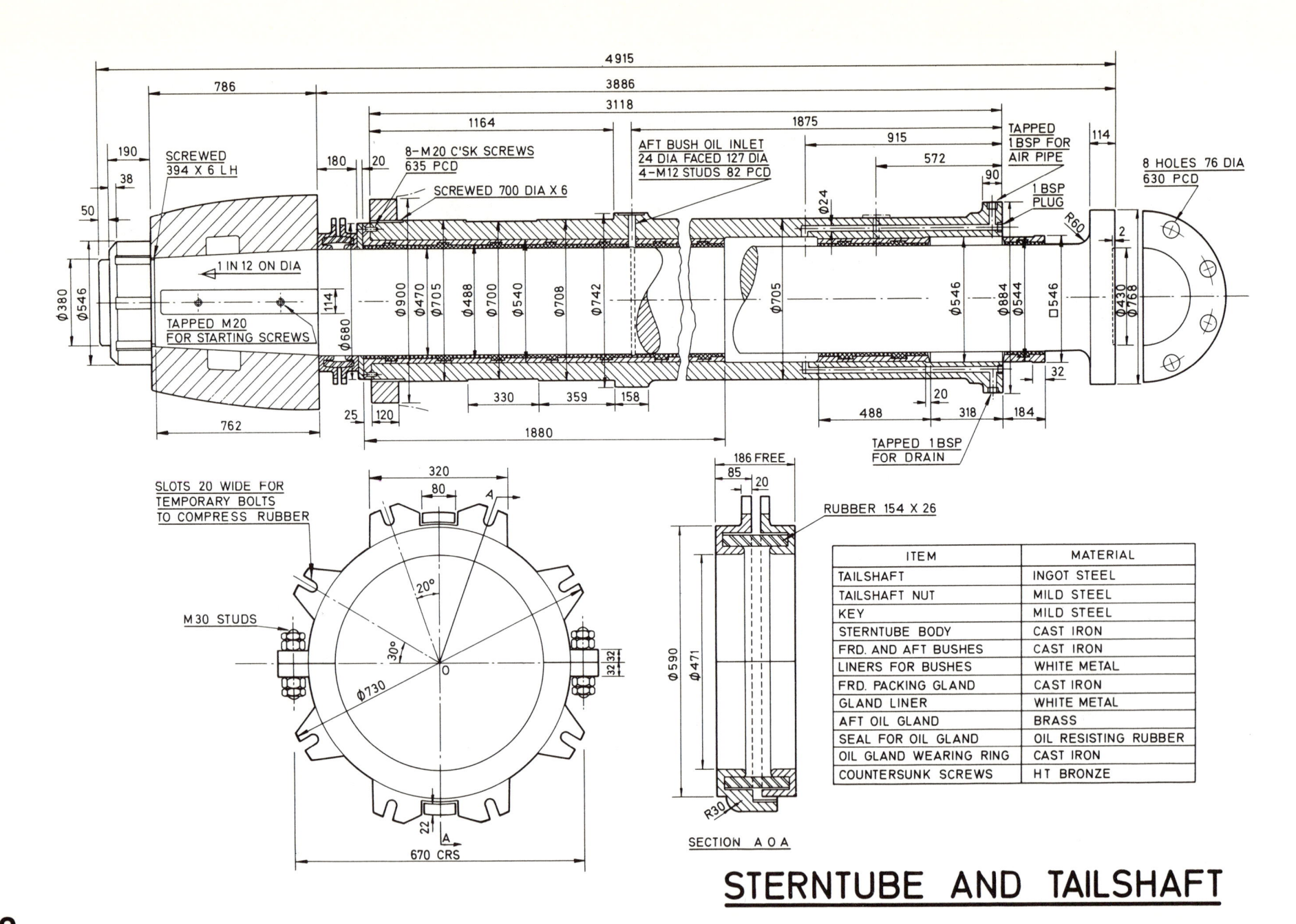

148

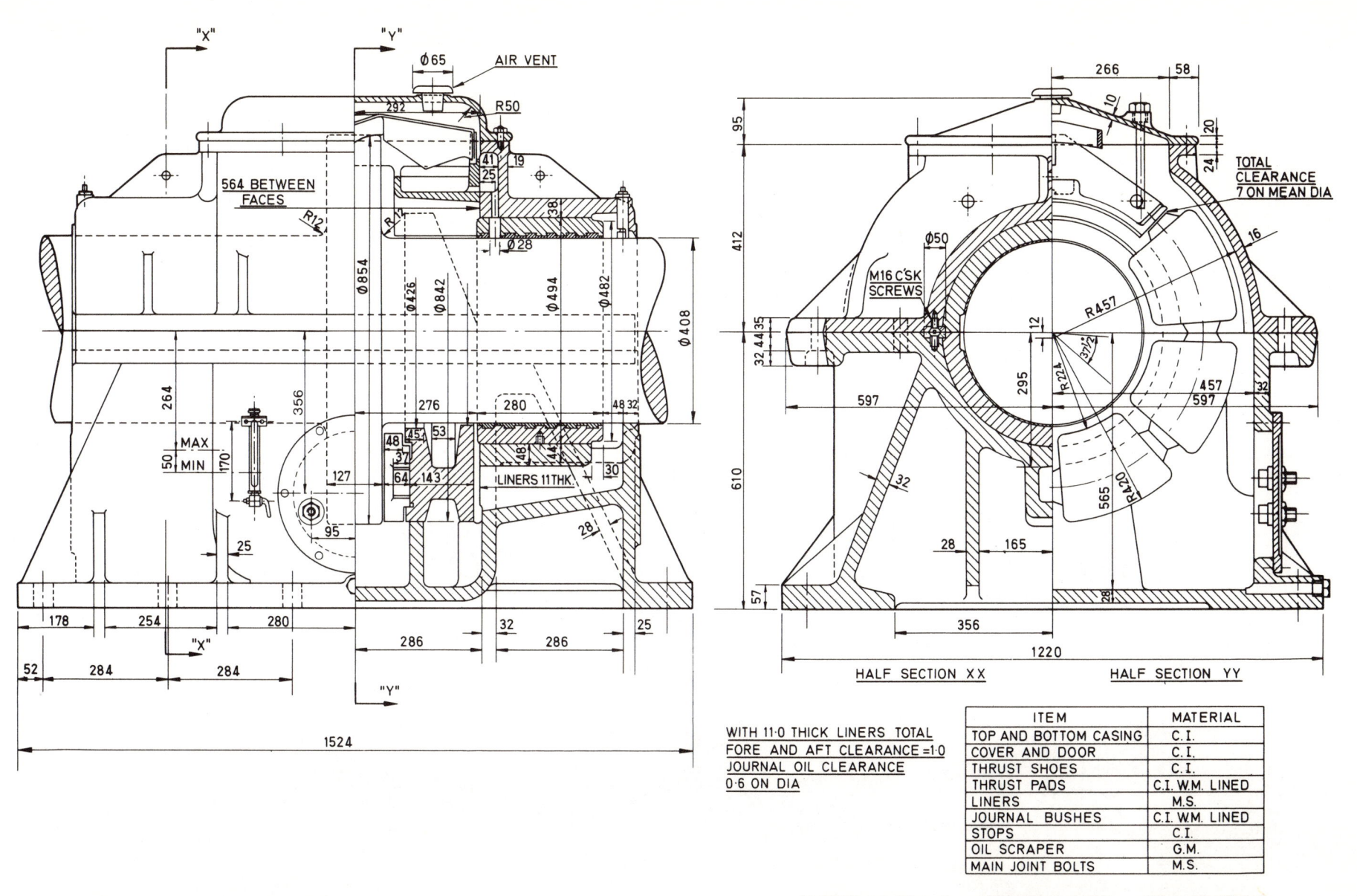

ITEM	MATERIAL
TOP AND BOTTOM CASING	C.I.
COVER AND DOOR	C.I.
THRUST SHOES	C.I.
THRUST PADS	C.I. W.M. LINED
LINERS	M.S.
JOURNAL BUSHES	C.I. W.M. LINED
STOPS	C.I.
OIL SCRAPER	G.M.
MAIN JOINT BOLTS	M.S.

WITH 11.0 THICK LINERS TOTAL FORE AND AFT CLEARANCE = 1.0
JOURNAL OIL CLEARANCE 0.6 ON DIA

MICHELL THRUST BLOCK

149

TURBINE MAIN GEAR WHEEL

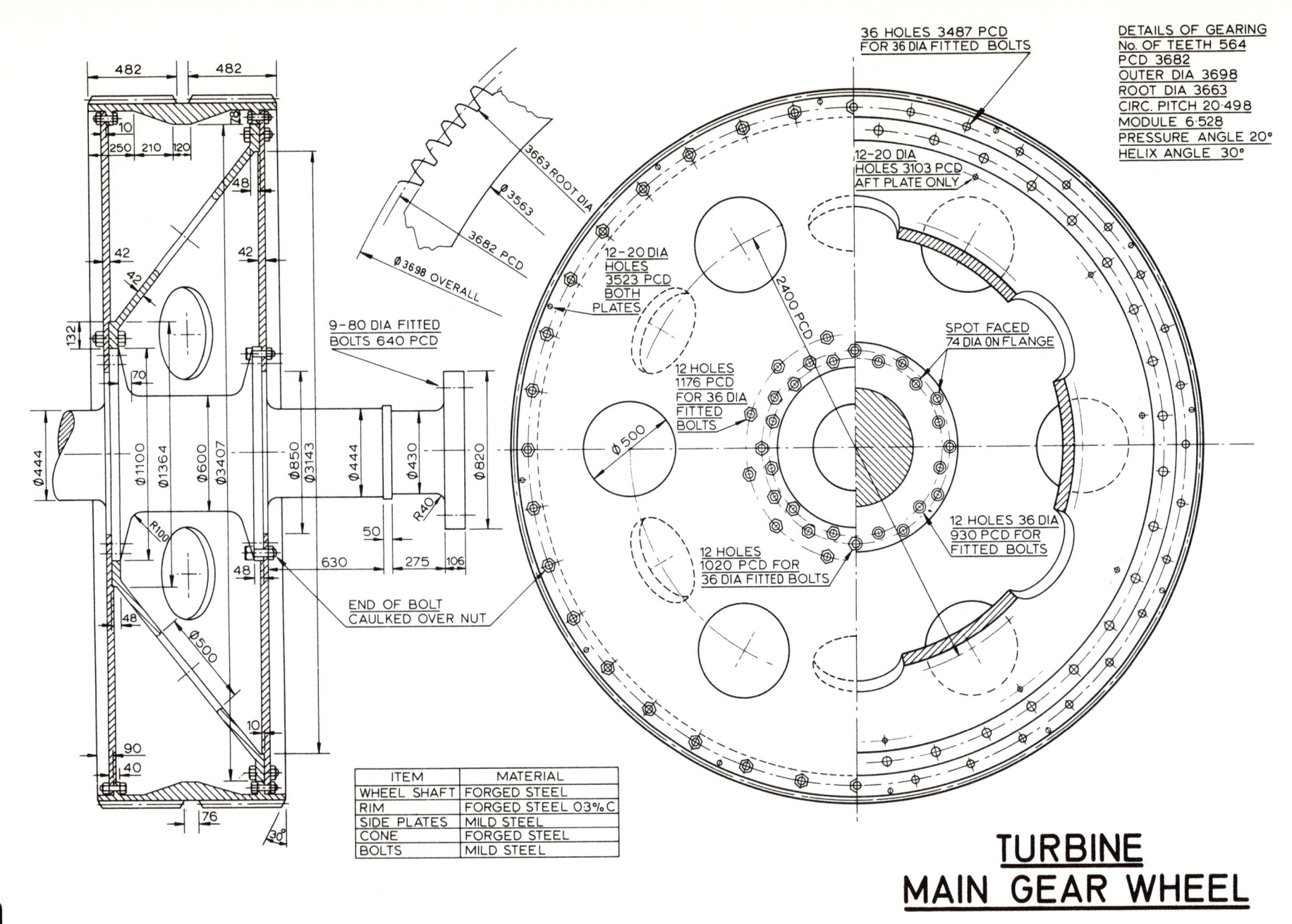

DETAILS OF GEARING
No. OF TEETH 564
PCD 3682
OUTER DIA 3698
ROOT DIA 3663
CIRC. PITCH 20·498
MODULE 6·528
PRESSURE ANGLE 20°
HELIX ANGLE 30°

ITEM	MATERIAL
WHEEL SHAFT	FORGED STEEL
RIM	FORGED STEEL 0·3%C
SIDE PLATES	MILD STEEL
CONE	FORGED STEEL
BOLTS	MILD STEEL

36 HOLES 3487 PCD FOR 36 DIA FITTED BOLTS
12-20 DIA HOLES 3103 PCD AFT PLATE ONLY
12-20 DIA HOLES 3523 PCD BOTH PLATES
12 HOLES 1176 PCD FOR 36 DIA FITTED BOLTS
SPOT FACED 74 DIA ON FLANGE
2400 PCD
12 HOLES 36 DIA 930 PCD FOR FITTED BOLTS
12 HOLES 1020 PCD FOR 36 DIA FITTED BOLTS
9-80 DIA FITTED BOLTS 640 PCD
END OF BOLT CAULKED OVER NUT

3663 ROOT DIA
Ø3563
Ø3682 PCD
Ø3698 OVERALL

150

HYDRAULIC STEERING GEAR

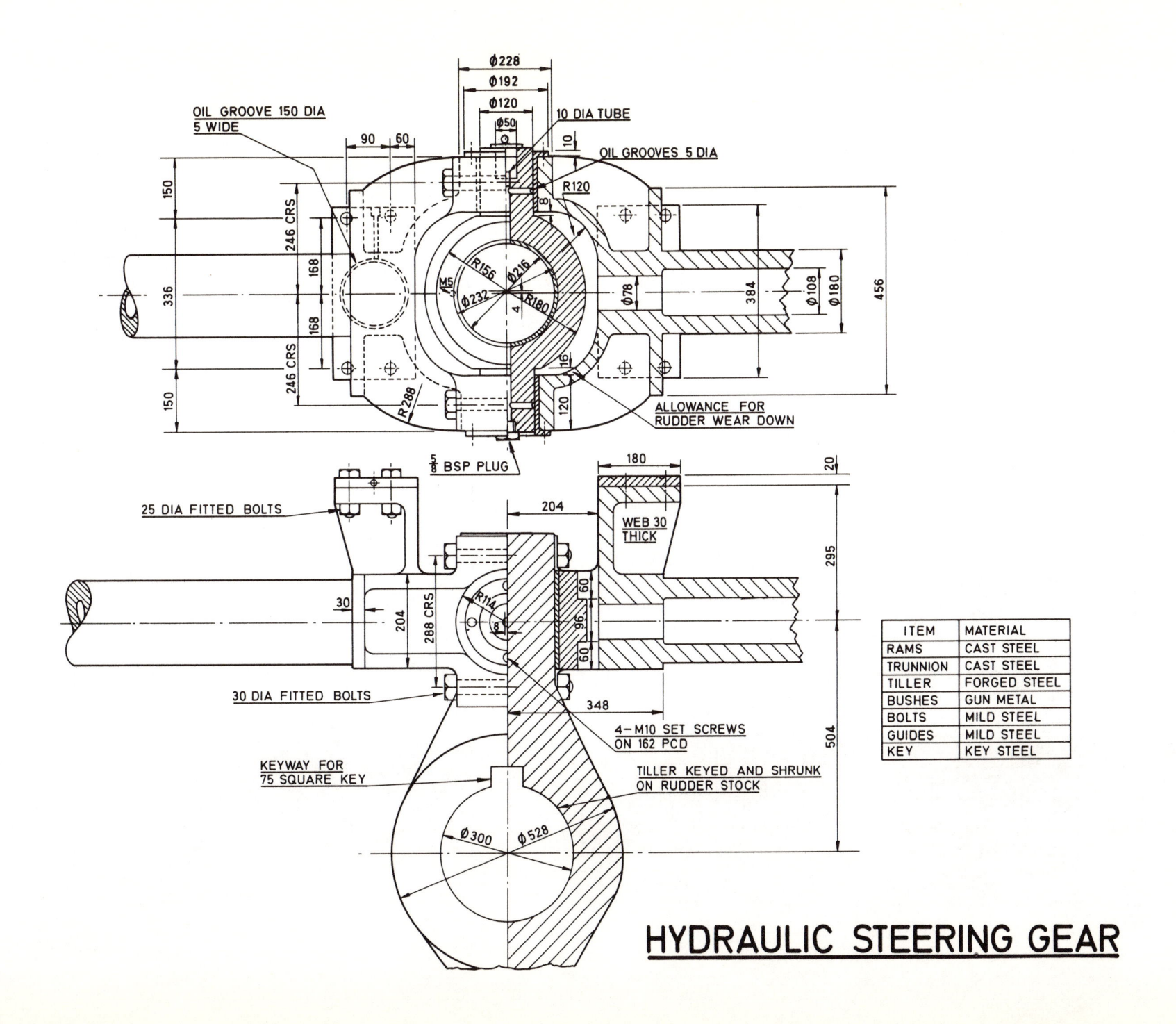

COMPRESSOR PISTON AND SUCTION VALVE

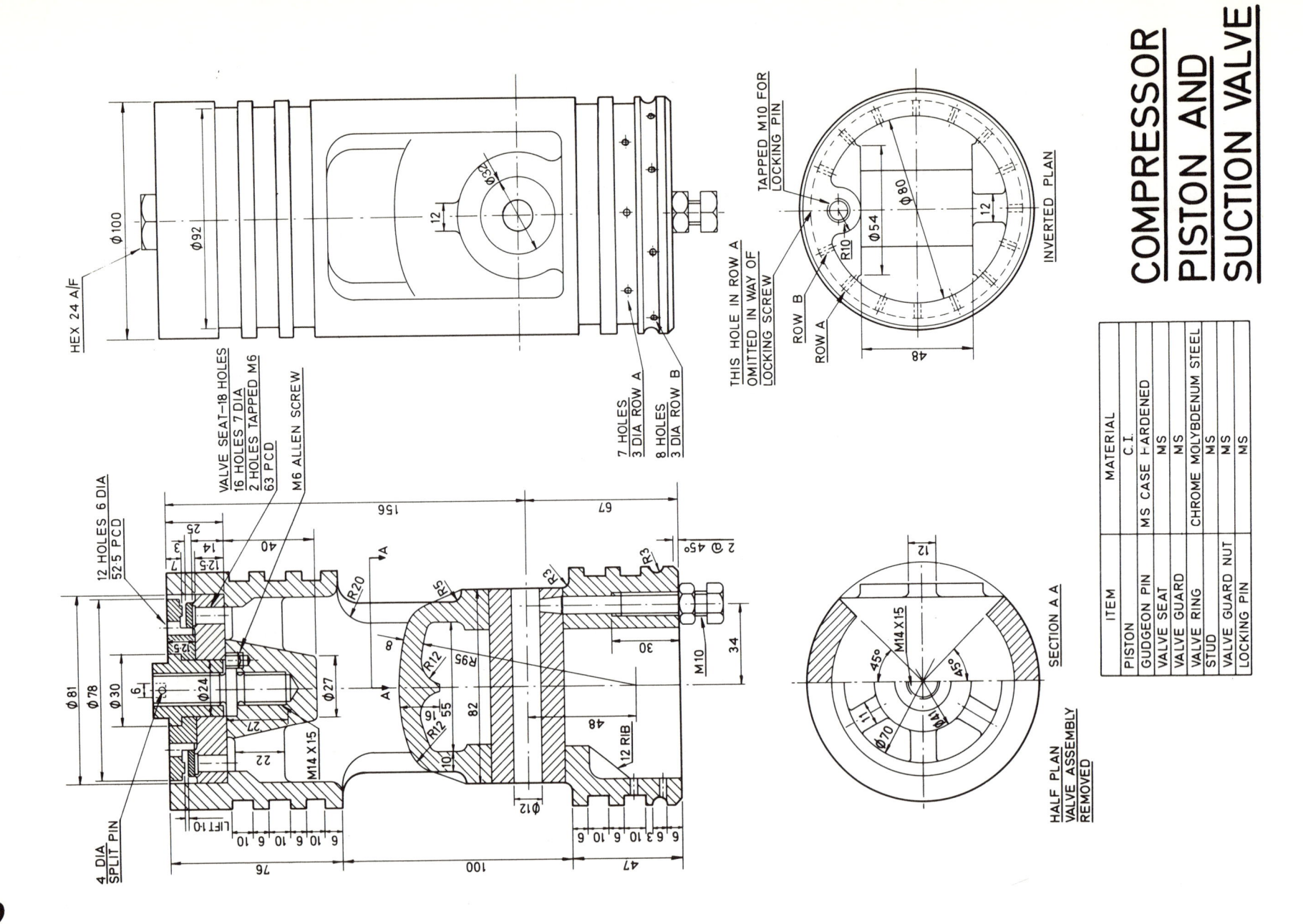

ITEM	MATERIAL
PISTON	C.I.
GUDGEON PIN	MS CASE HARDENED
VALVE SEAT	MS
VALVE GUARD	MS
VALVE RING	CHROME MOLYBDENUM STEEL
STUD	MS
VALVE GUARD NUT	MS
LOCKING PIN	MS

PLATE TYPE GAUGE GLASS

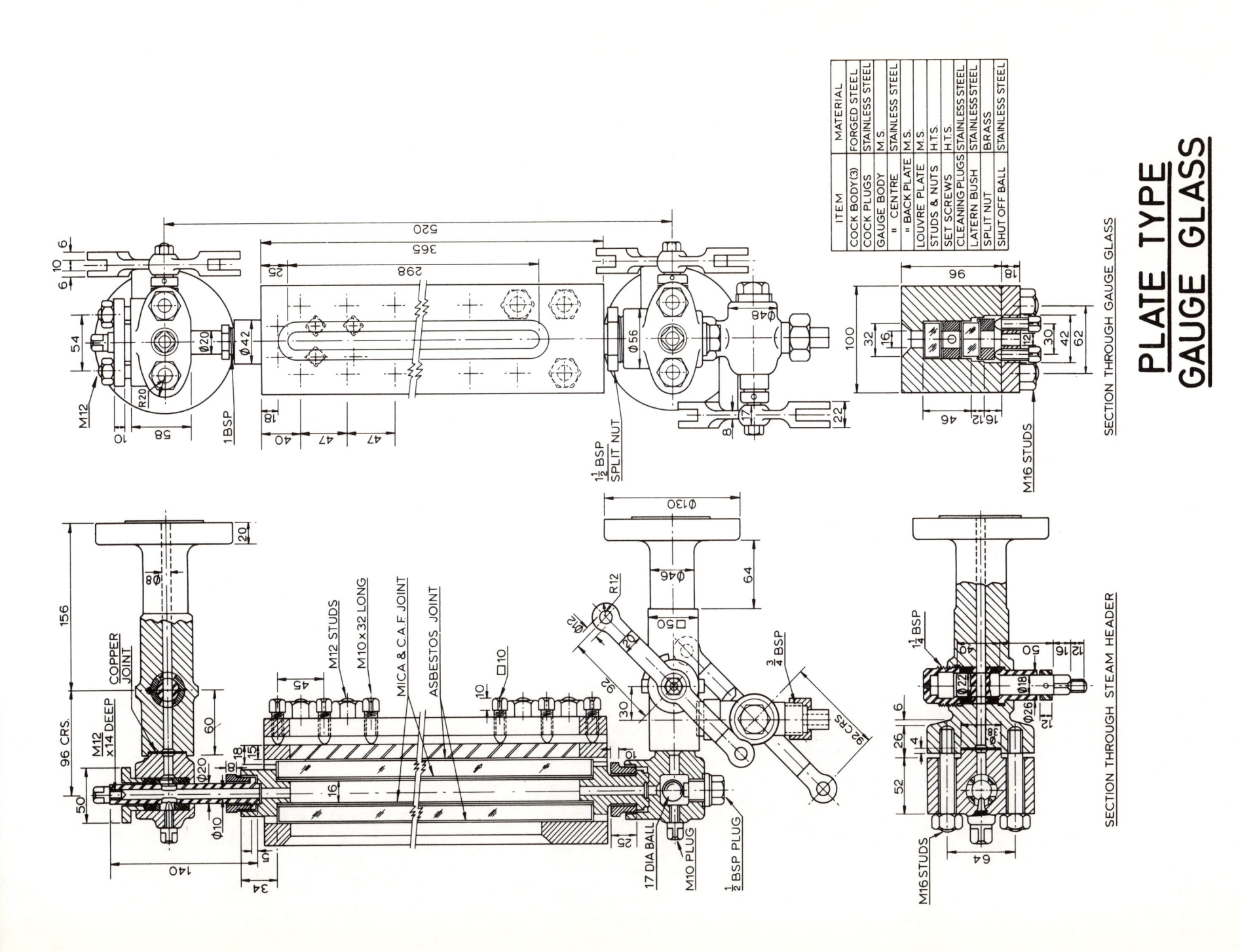

ITEM	MATERIAL
COCK BODY (3)	FORGED STEEL
COCK PLUGS	STAINLESS STEEL
GAUGE BODY	M.S.
" CENTRE	STAINLESS STEEL
" BACK PLATE	M.S.
LOUVRE PLATE	M.S.
STUDS & NUTS	H.T.S.
SET SCREWS	STAINLESS STEEL
CLEANING PLUGS	STAINLESS STEEL
LATERN BUSH	STAINLESS STEEL
SPLIT NUT	BRASS
SHUT OFF BALL	STAINLESS STEEL

SECTION THROUGH GAUGE GLASS

SECTION THROUGH STEAM HEADER

153

MECHANICAL LUBRICATOR

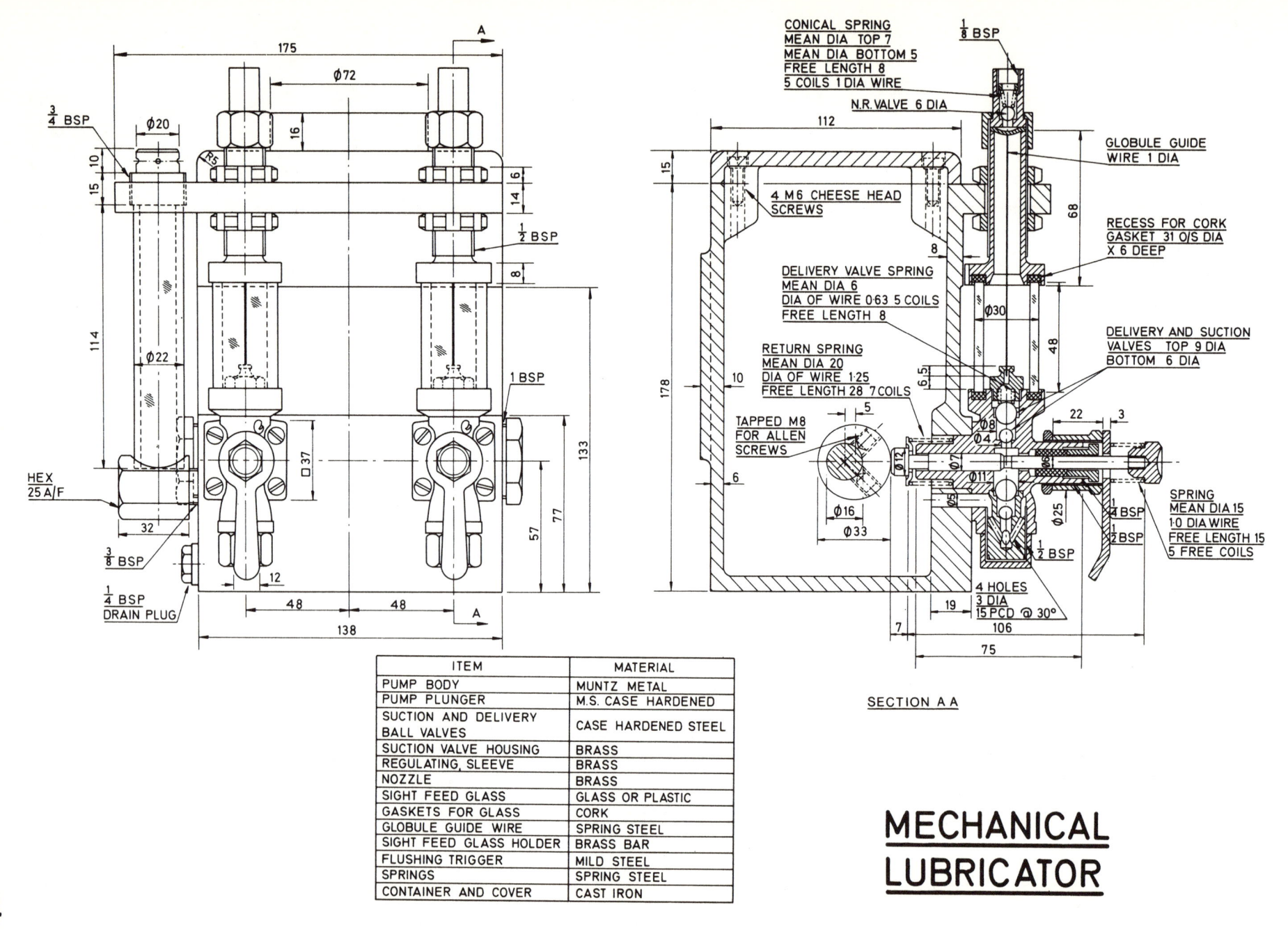

ITEM	MATERIAL
PUMP BODY	MUNTZ METAL
PUMP PLUNGER	M.S. CASE HARDENED
SUCTION AND DELIVERY BALL VALVES	CASE HARDENED STEEL
SUCTION VALVE HOUSING	BRASS
REGULATING SLEEVE	BRASS
NOZZLE	BRASS
SIGHT FEED GLASS	GLASS OR PLASTIC
GASKETS FOR GLASS	CORK
GLOBULE GUIDE WIRE	SPRING STEEL
SIGHT FEED GLASS HOLDER	BRASS BAR
FLUSHING TRIGGER	MILD STEEL
SPRINGS	SPRING STEEL
CONTAINER AND COVER	CAST IRON

SECTION A A

FIRST ANGLE SYMBOL

THIRD ANGLE SYMBOL

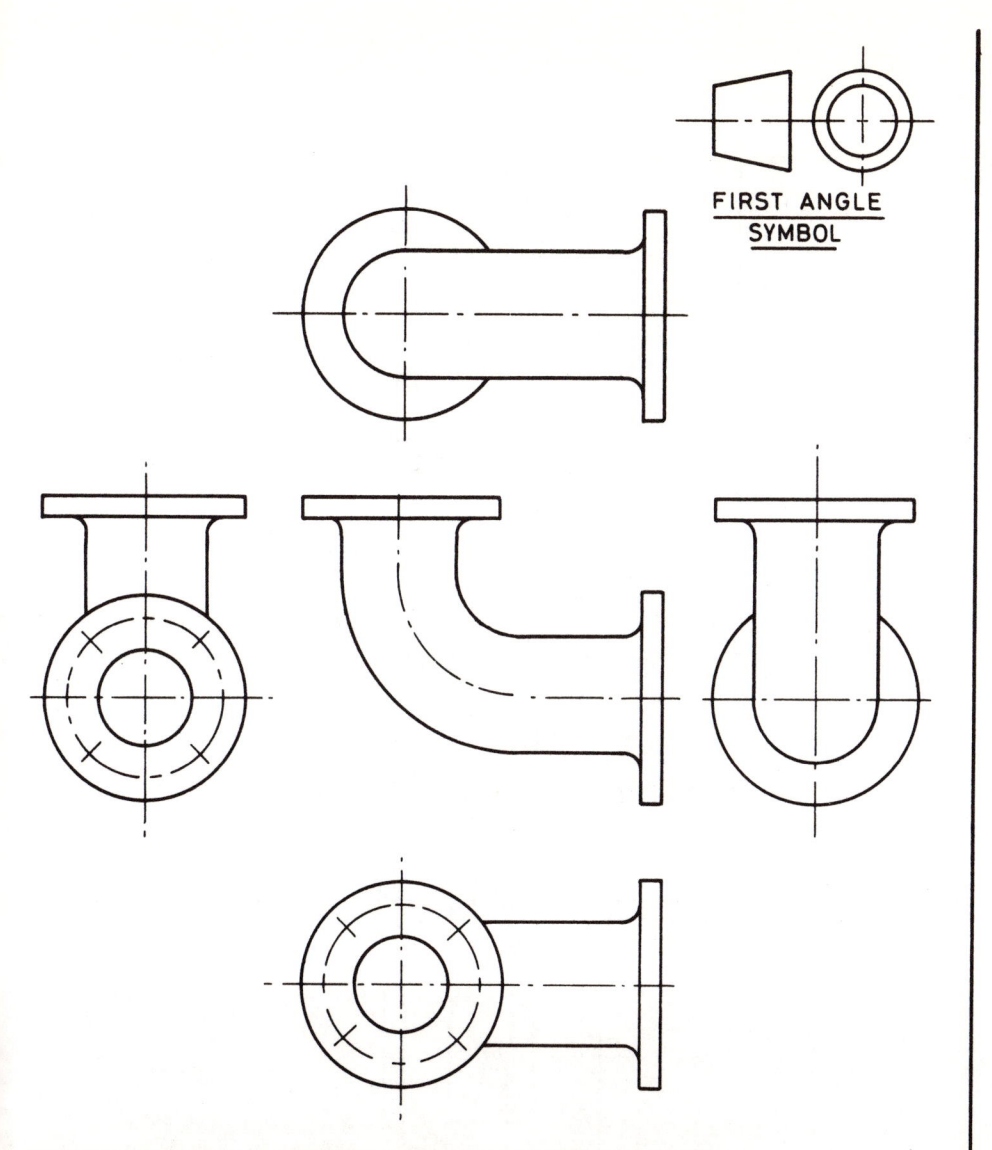

PIPE BEND IN FIRST ANGLE PROJECTION
IN FIRST ANGLE PROJECTION EACH VIEW SHOWS WHAT WOULD BE SEEN BY LOOKING ON THE FAR SIDE OF AN ADJACENT VIEW

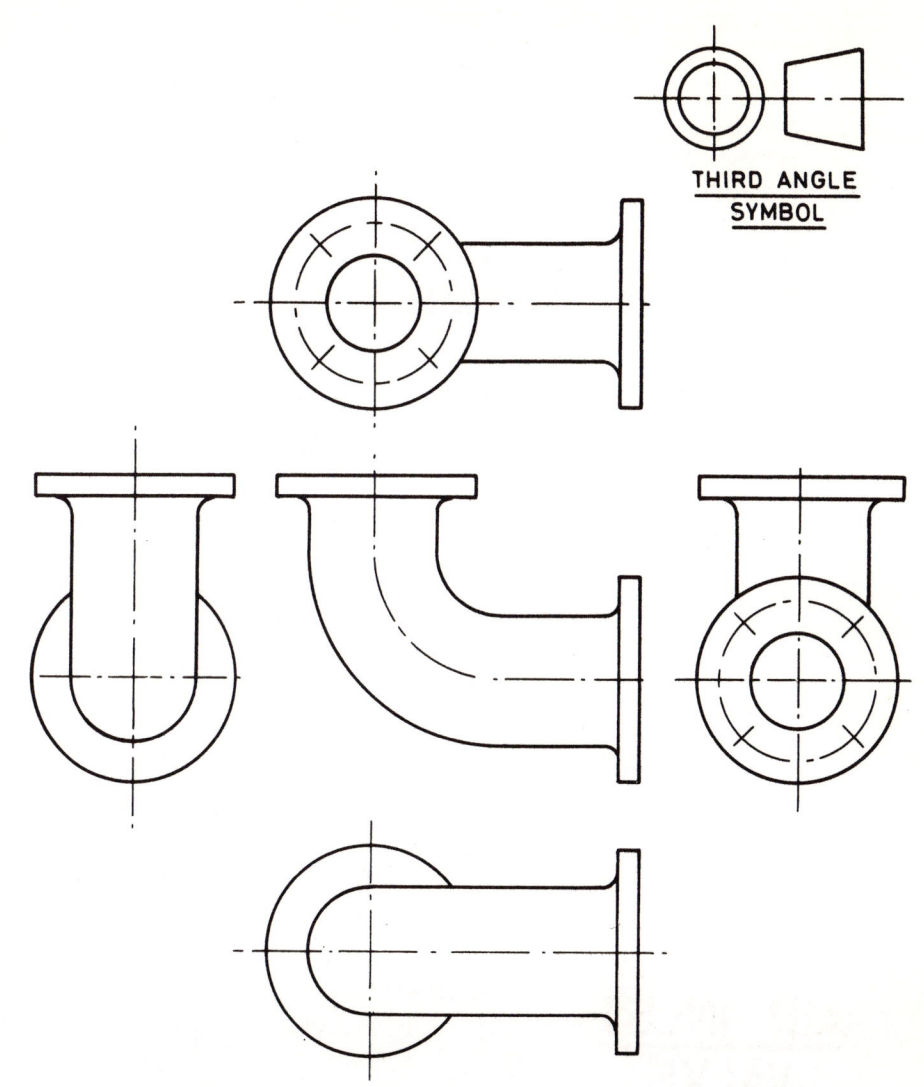

PIPE BEND IN THIRD ANGLE PROJECTION
IN THIRD ANGLE PROJECTION EACH VIEW SHOWS WHAT WOULD BE SEEN BY LOOKING ON THE NEAR SIDE OF AN ADJACENT VIEW

AIR INLET VALVE

PART No.	ITEM	No. OFF	MATERIAL
1	VALVE BODY	1	CAST IRON
2	VALVE SPINDLE	1	HEAT RESISTING STEEL EN 55
3	PISTON	1	MILD STEEL
4	SPRING KEEP	1	" "
5	SPLIT COLLAR	1	" "
6	SPLIT RING	1	" "
7	PISTON RINGS	3	CAST IRON
8 9	SPRINGS	2	SPRING STEEL

SPRING	MEAN DIA	WIRE DIA	No. COILS	FREE LENGTH	REMARKS
INNER	200	10	7	325	R H HELIX
OUTER	260	15	5·5	325	L H HELIX

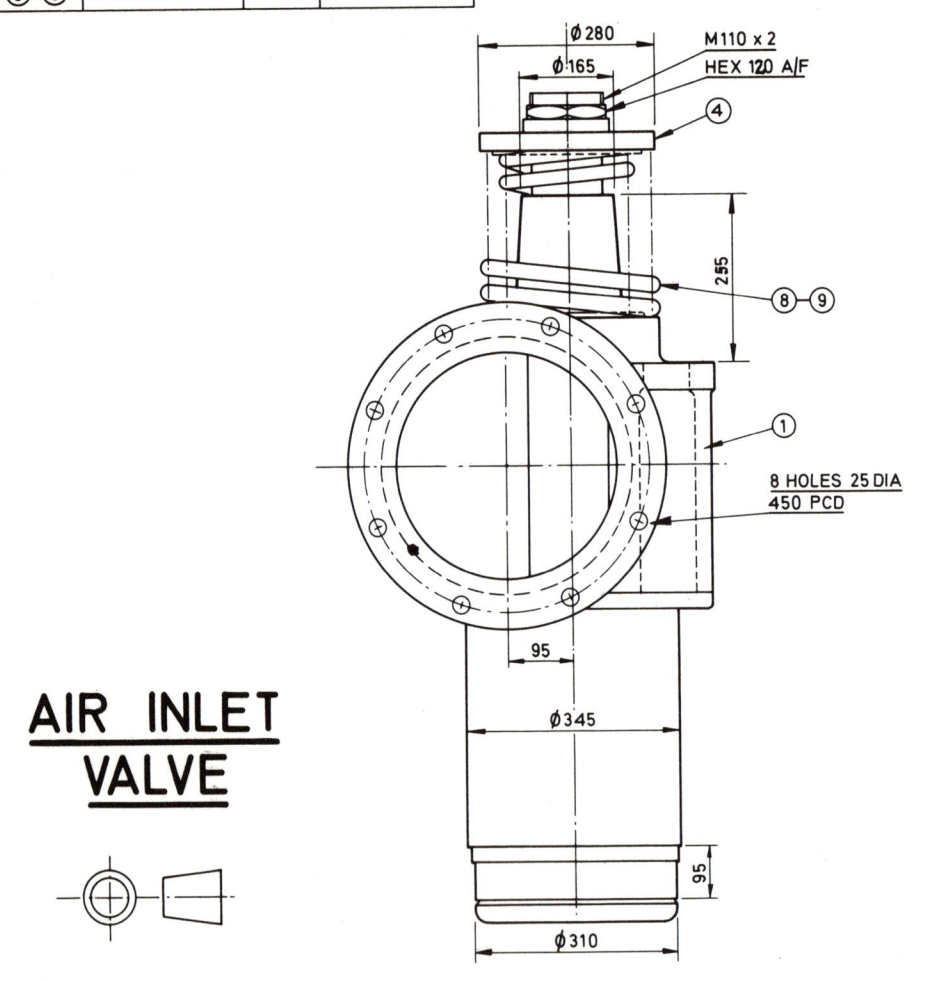

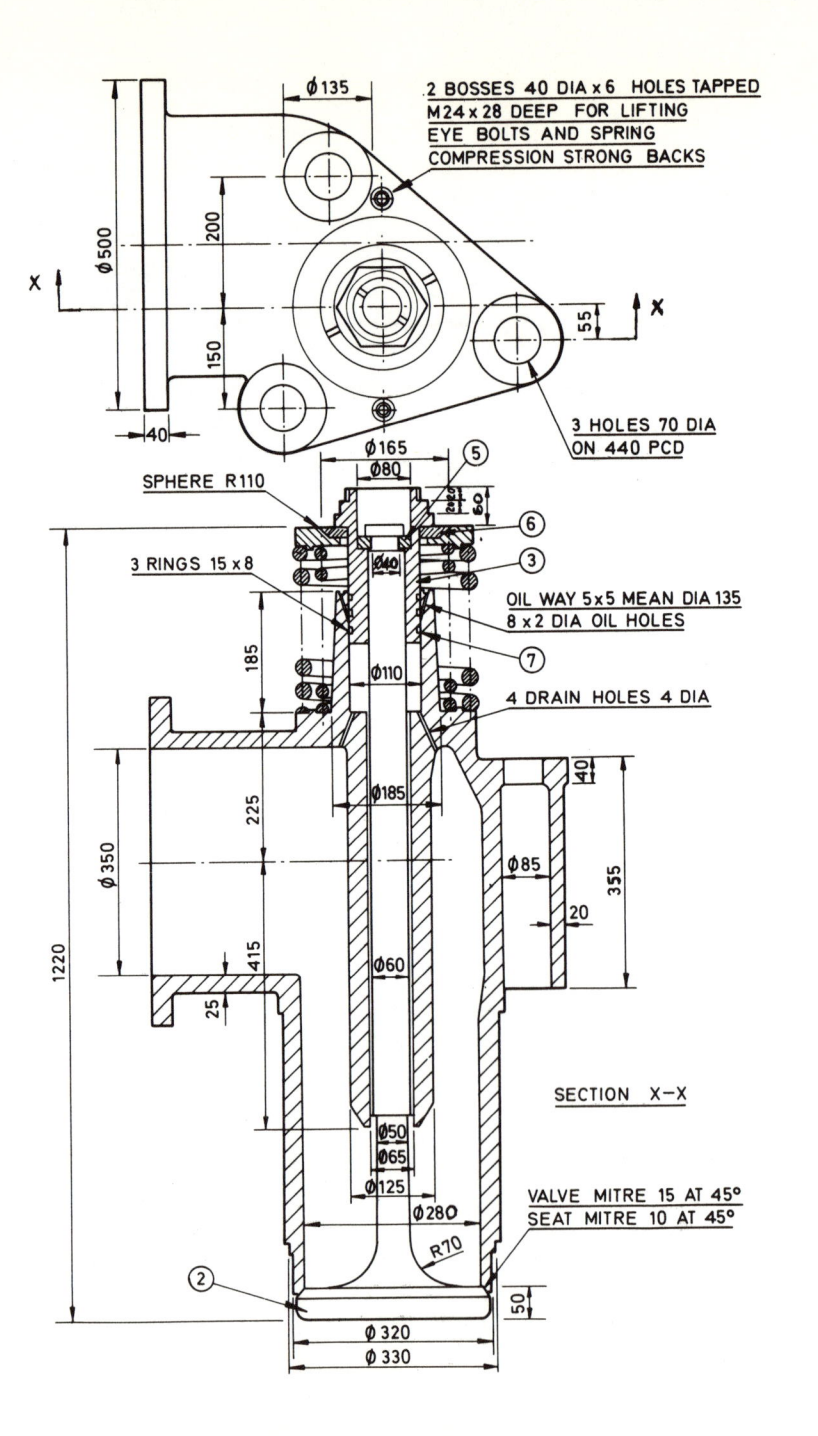

Chapter Four

STATICS

When investigating problems connected with forces, it will be found convenient to represent a force completely by a straight line. For this purpose the straight line must show:—

(1) The direction of the line of action of the force in relation to some fixed direction.
(2) The magnitude of the force shown by the length of the line, measured on a suitable scale.
(3) The sense in which the force acts along the line of action, indicated by an arrowhead.

When such a straight line is used to represent a force it is known as a vector, and the geometrical figure constructed from details given in the space diagram, is known as a vector diagram.

In solving problems, either of two methods can be used—a graphical solution, or a solution by calculation. For the latter method a working knowledge of trigonometry is essential, and those students who are weak in mathematics at this stage, will need to use the graphical method. It should be emphasised that the graphical method is not suggested in preference to the mathematical treatment, and students are advised to study the mathematics involved, clearly set out in a companion volume 'Reeds Applied Mechanics for Engineers'.

In the graphical method, the vector diagram is drawn to scale, and the unknown quantity measured off the diagram, the magnitude by a scale rule, and the direction by a protractor. The accuracy of the drawing determines the accuracy of the answer, so vector diagrams should be drawn to the largest scale possible.

Parallelogram of forces

If two forces acting on a body can be represented by two adjacent sides of a parallelogram drawn from the point of application, their resultant will be the diagonal of the parallelogram drawn from that point. The resultant of two or more forces is that single force which can replace them and have the same effect. Another term in use is the Equilibrant, which could be defined as that force which is equal in magnitude to the resultant, but opposite in sense.

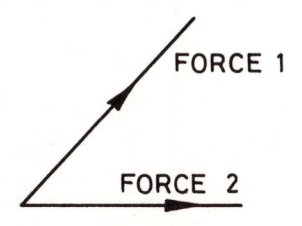

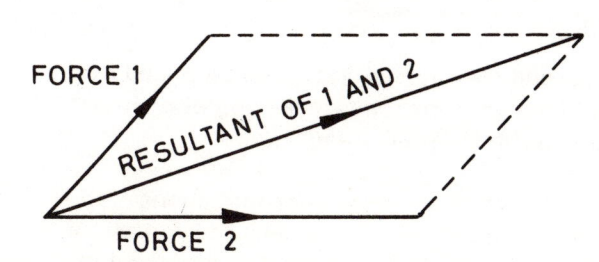

Triangle of Forces

If three forces acting at a point are in equilibrium, the vector diagram drawn to scale representing the forces in magnitude and direction, taken in order, forms a closed triangle.

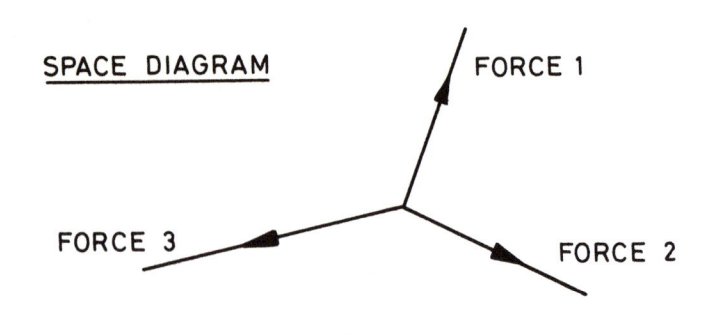

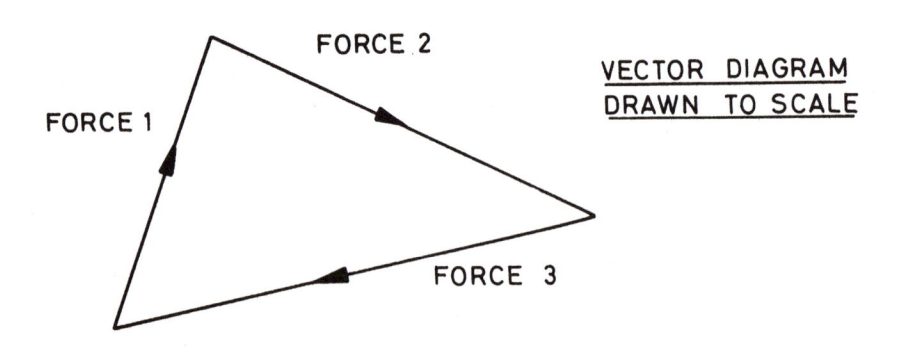

SPACE DIAGRAM

FORCE 1
FORCE 2
FORCE 3

VECTOR DIAGRAM DRAWN TO SCALE

FORCE 1
FORCE 2
FORCE 3

Example

Two ropes are slung from a beam, and their lower ends are connected by a shackle from which a load of 500 N hangs. If the ropes make angles of 45 and 70 degrees respectively to the vertical, find the pull in each rope.

Commence by drawing the space diagram to illustrate the connections of the ropes and the load and insert the letters A, B, C between the forces. To construct the vector diagram, draw a b vertically downwards to represent the force AB to some convenient scale. From b draw a line parallel to the force between spaces B and C, and from a draw a line parallel to the force between the spaces C and A. This method of lettering the spaces of the space diagram with capital letters, so that each force can be referred to by the letters of the two spaces the force separates, and the vector diagram labelled with the corresponding small letter on the ends of the vector in the direction of the arrow, is known as BOW'S NOTATION. The triangle thus formed is the vector diagram for the configuration, and the tension in the ropes can be found by measuring the lengths bc and ac to the scale of the diagram.

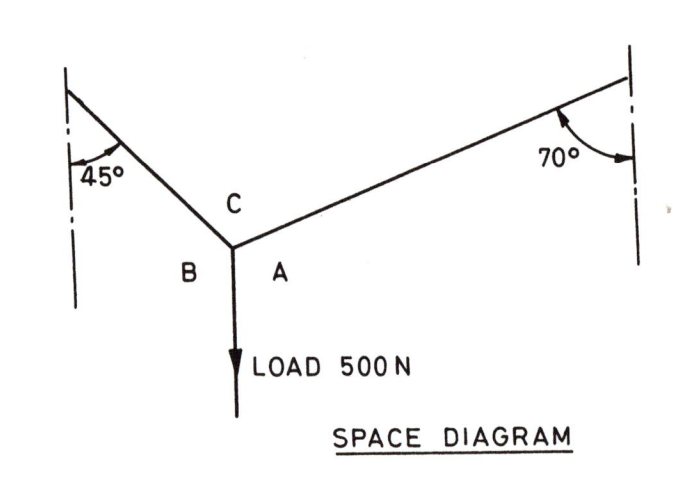

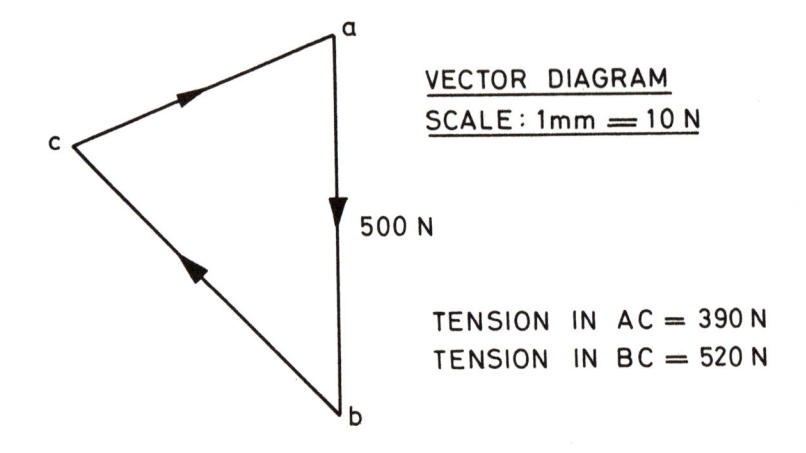

SPACE DIAGRAM

45° 70°
C
B A
LOAD 500 N

VECTOR DIAGRAM
SCALE: 1mm = 10 N

500 N

TENSION IN AC = 390 N
TENSION IN BC = 520 N

Example

A simple jib crane supports a mass of 100kg and the lengths of the post, jib and the tie are 2½m, 4½m and 3½m respectively. Determine the magnitude of the forces in the jib and tie.

To construct the vector diagram, draw ab vertical to represent 981N to scale.

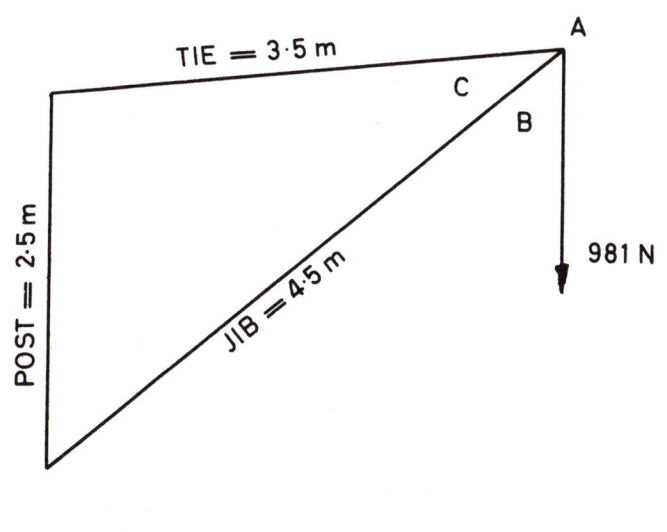

SPACE DIAGRAM
SCALE: 20mm = 1m

VECTOR DIAGRAM
SCALE: 1mm = 15 N

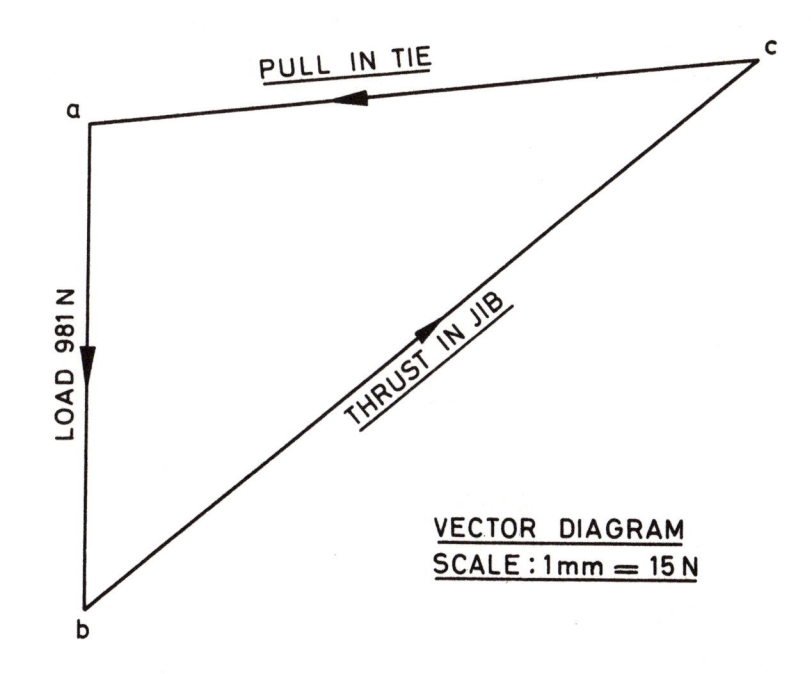

Before commencing to draw the vector diagram the mass of 100kg will have to be converted into a force equal to the gravitational pull on the mass.

Mass of load = 100kg

Force in rope to overcome gravitational force i.e. the weight =

$$100 \times 9.81 = 981N.$$

From a draw ac parallel to the force in the space AC, and from b draw bc parallel to the force between B and C. In the triangle so formed, measure ac and bc to determine the forces in the tie and jib respectively.

ANSWER: Pull in tie = 1374N
Thrust in jib = 1770N

159

Polygon of Forces

If any number of forces acting at a point are in equilibrium, the vector diagram drawn to scale representing the forces in magnitude and direction, taken in order, forms a closed polygon. This theorem is similar to the Triangle of Forces, except that where the Triangle of Forces refers only to three forces, the Polygon of Forces refers to any number greater than three.

System of Forces Not in Equilibrium

If a system of forces acting about a point is given, it may not be in a state of equilibrium. The following example shows how the force necessary to produce equilibrium can be obtained from the vector diagram.

Example

The following forces act about a point.

50N due North
60N at 10° North of East
50N South East
40N at 20° West of South

Draw a vector diagram and determine the magnitude and direction of the equilibrant.

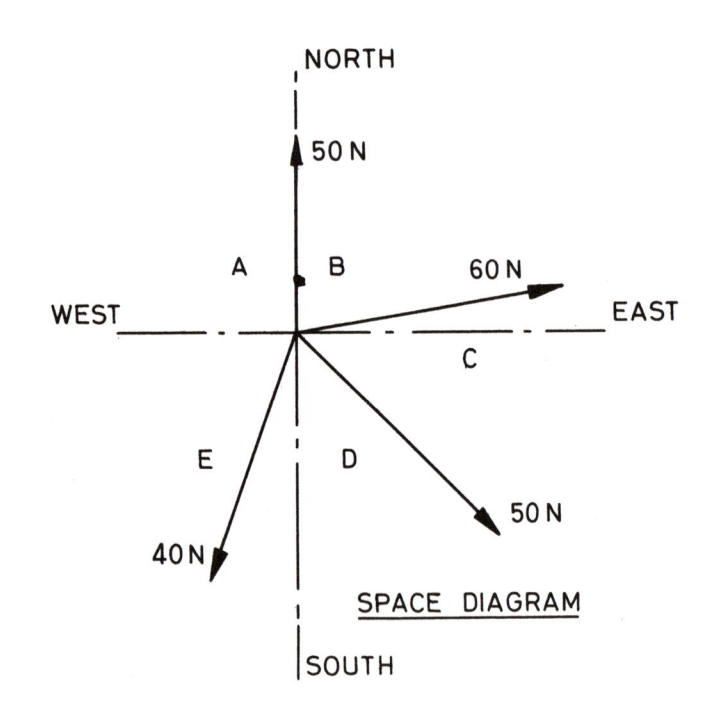

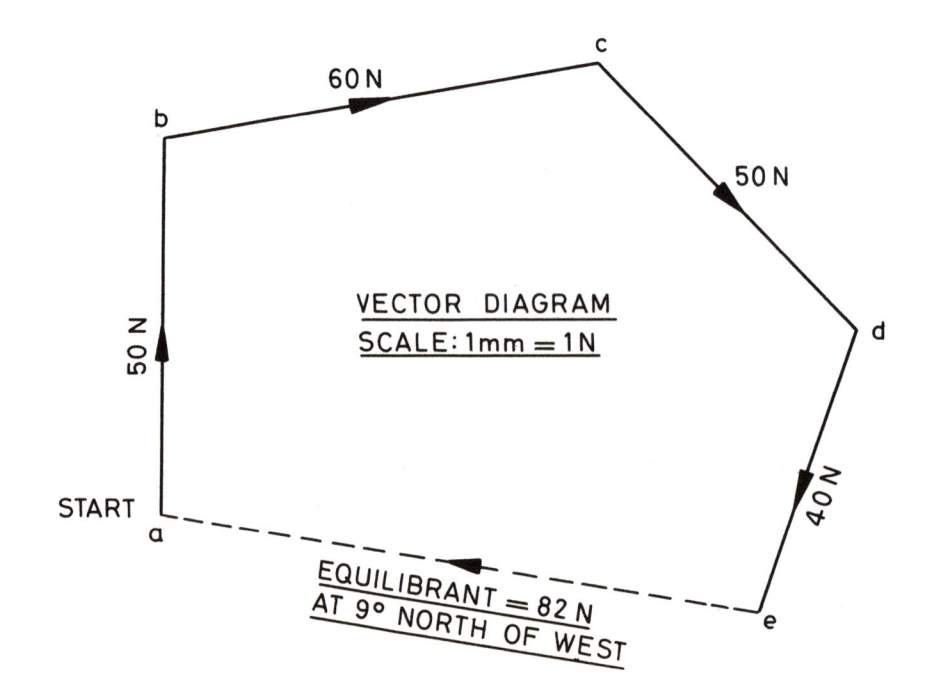

SPACE DIAGRAM

VECTOR DIAGRAM
SCALE: 1mm = 1N

EQUILIBRANT = 82 N AT 9° NORTH OF WEST

160

Reciprocating Engine Mechanism

The connecting rod and crank of a reciprocating engine converts the reciprocating motion of the piston into a rotary motion at the crank shaft.

Considering the forces meeting at the crosshead, the piston rod pushes vertically downward on the crosshead, the thrust in the connecting rod appears as an upward resisting force inclined to the vertical, and the guide exerts a horizontal force to balance the horizontal component of the thrust in the connecting rod. As the piston effort always acts vertically, and the guide force always horizontally, the vector diagram of the forces at the crosshead is always a right-angled triangle.

Example

The piston of a reciprocating engine exerts a force of 150 kN on the crosshead when the crank is 65° past top dead centre. If the stroke of the engine is 900mm, and the length of the connecting rod is 1.6m, find the guide force and the force in the connecting rod.

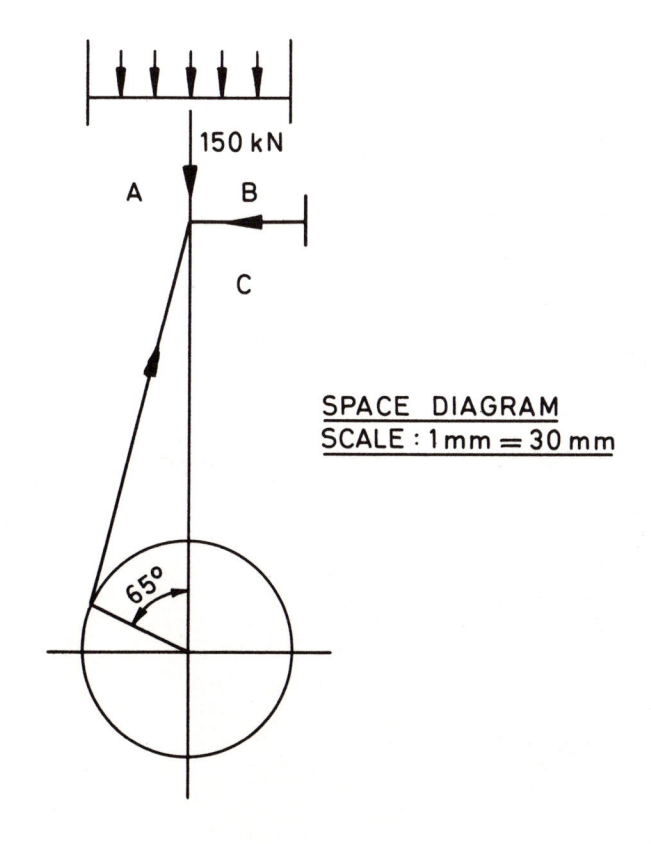

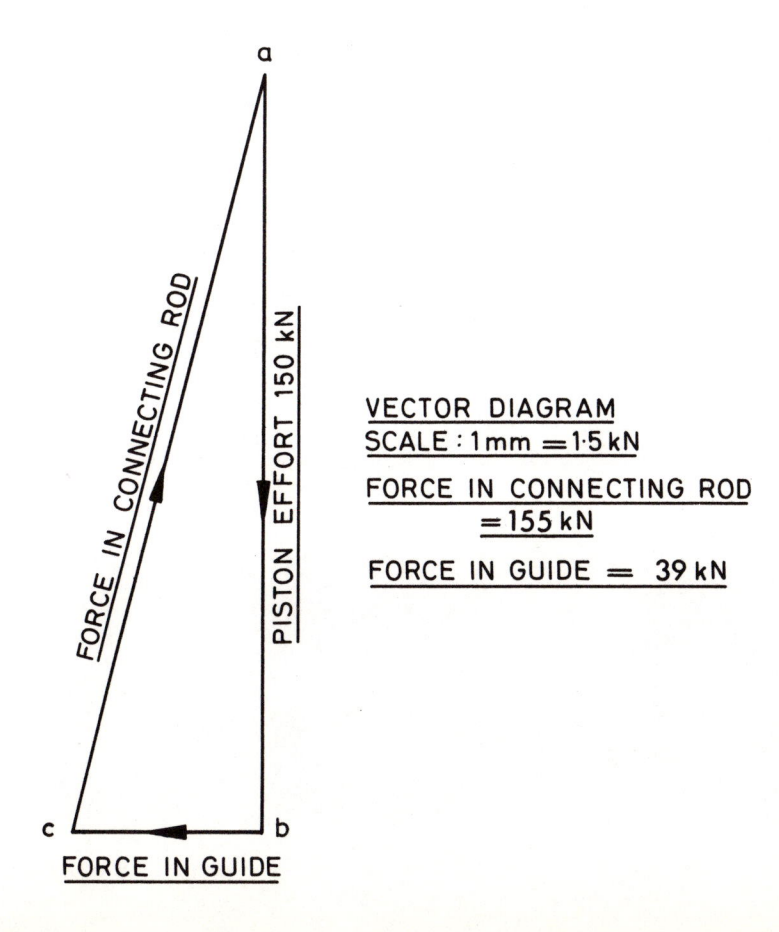

SPACE DIAGRAM
SCALE : 1mm = 30 mm

VECTOR DIAGRAM
SCALE : 1mm = 1·5 kN

FORCE IN CONNECTING ROD = 155 kN

FORCE IN GUIDE = 39 kN

The following example illustrates how velocities can be represented on a vector diagram. A ship travelling due North at 20 knots runs into a 4 knot current moving South East. By drawing a vector diagram find the resultant speed and direction of the ship.

Draw in space diagram as shown, and in the vector diagram draw ab parallel to AB to a suitable scale. Draw bc parallel to BC and also to scale. The line closing the triangle gives the resultant velocity in magnitude and direction.

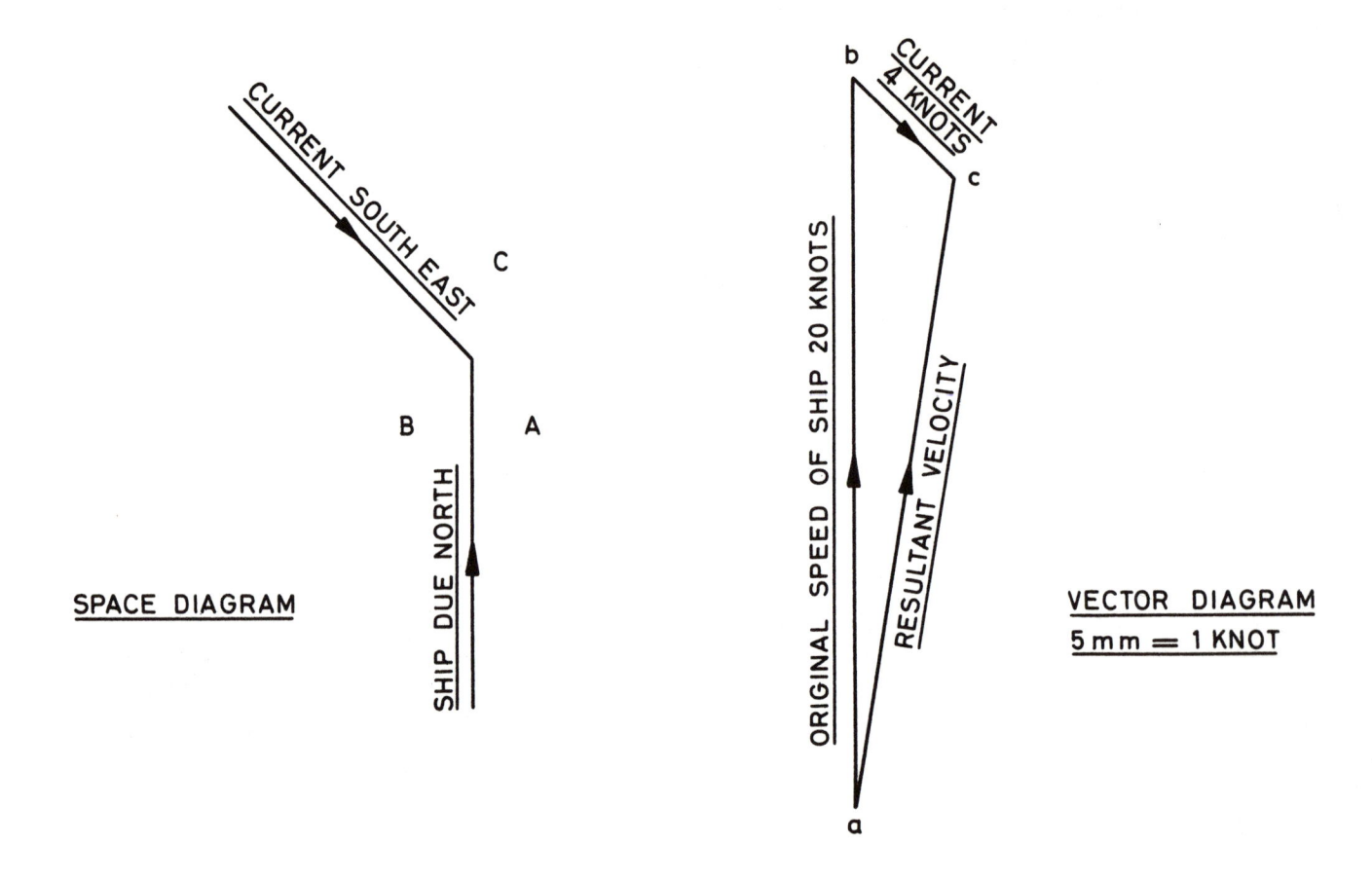

SPACE DIAGRAM

VECTOR DIAGRAM
5 mm = 1 KNOT

162

Ans. 17.5 knots at 9½° East of North.

Shearing Force and Bending Moment Diagrams

In the design of beams, it is often required to know the maximum shearing force and bending moment on the beam, and also values for these at different sections along the length of the beam. To obtain these values, graphs are drawn to represent the shearing force and bending moment variations over the length of the beam, and these diagrams are known as Shearing Force and Bending Moment diagrams. In drawing these diagrams, the graphs may be plotted either above or below the base line, and they should be drawn to some scale such as 50mm = 1m of beam length, 10mm = 1 kilonewton of shearing force, and 5mm = 1 kilonewton-metre of bending moment. These figures of course would be chosen to suit the problem.

Some examples are given on the following pages of shearing force and bending moment diagrams for beams supported by different methods, and loaded in various ways, but for a complete treatise on the whole subject of shearing forces and bending moments, the student is advised to study the topic from a book on Applied Mechanics.

In the first few examples, the bending moments will be calculated for a number of points along the beam, so that the diagram can be plotted. If the shapes of the diagrams are observed it will soon be realised that they follow standard patterns, depending upon the type of loading, so that eventually it will only be necessary to find the values of the bending moment at certain points along the beam, and to join these points by either a straight line or a curve.

Example

A cantilever 5 metres long carries a concentrated load of 40 kN at the free end. Neglecting the weight of the beam, draw the shearing force and bending moment diagrams.

To construct the bending moment diagram, take moments at say every metre from the free end. Using M to represent the bending moment, M @ 1m = 40 kNm. M @ 2m = 40 x 2 = 80 kNm. M @ 3m = 40 x 3 = 120 kNm. M @ 4m = 40 x 4 = 160 kNm. M @ 5m = 40 x 5 = 200 kNm. On a base line of equal length to the

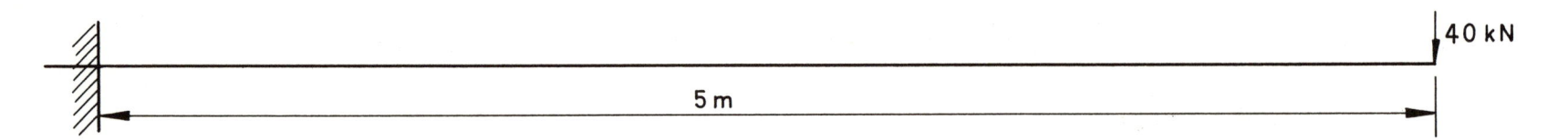

The shearing force diagram is a diagram of up and down forces. Draw the base line to a scale of say 1m beam length = 40mm. Starting at the free end there is a vertical downward force of 40 kN, therefore draw a line vertically downwards 20mm long, which is to the scale of 2 kN = 1mm. Moving toward the fixed end, there are no other forces on the beam until we get to the wall, so the graph is a horizontal line. At the wall there is a reaction of 40 kN vertically upwards, so a vertical line 20mm long at this point closes the diagram.

shearing force diagram, measure down at 1m intervals the bending moments calculated above, using a scale of say 4 kNm = 1mm. Join the points obtained from measuring down from the base line, and it will be found that a straight line is obtained, sloping from zero at the free end to maximum value at the fixed end.

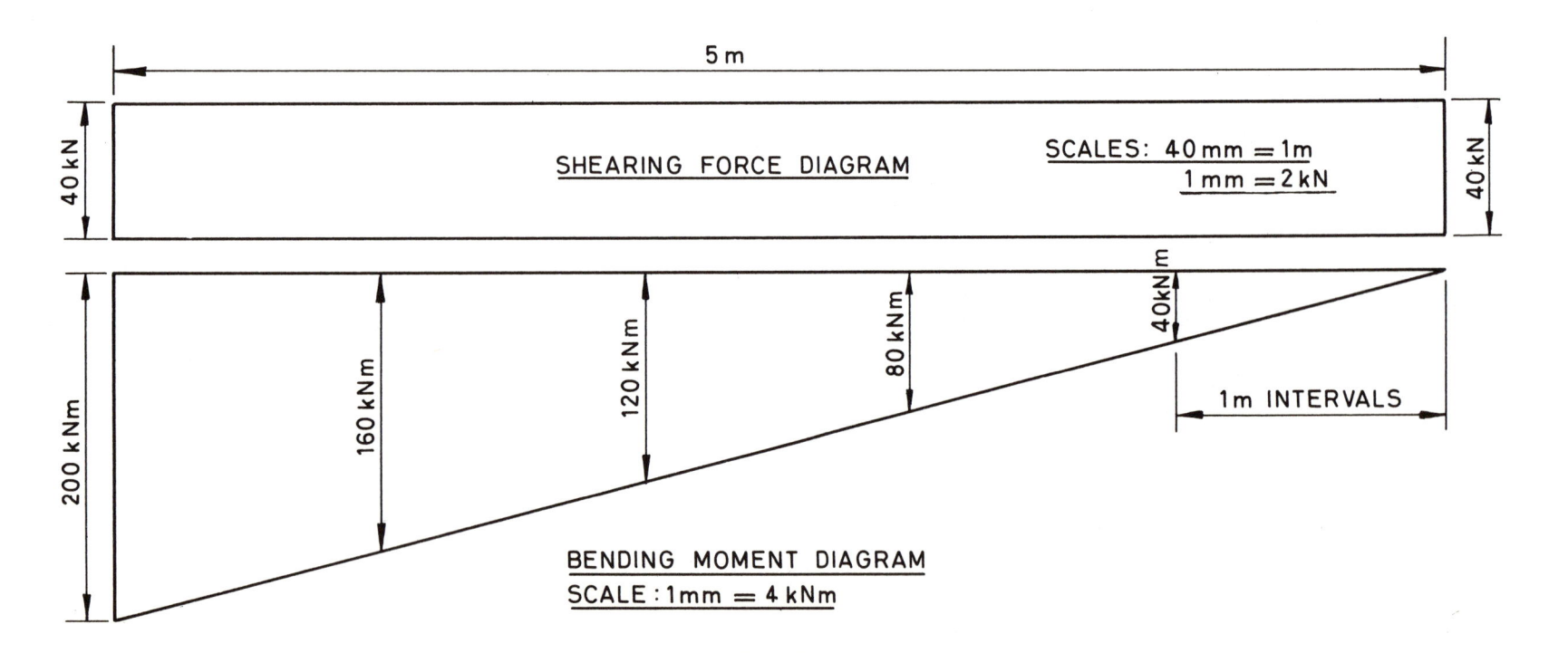

SHEARING FORCE DIAGRAM SCALES: 40 mm = 1m / 1 mm = 2 kN

BENDING MOMENT DIAGRAM
SCALE : 1mm = 4 kNm

Example

A cantilever five metres long carries a uniformly distributed load over its entire length, which together with the weight of the beam amounts to 30kN per metre run. Draw the shearing force and bending moment diagrams.

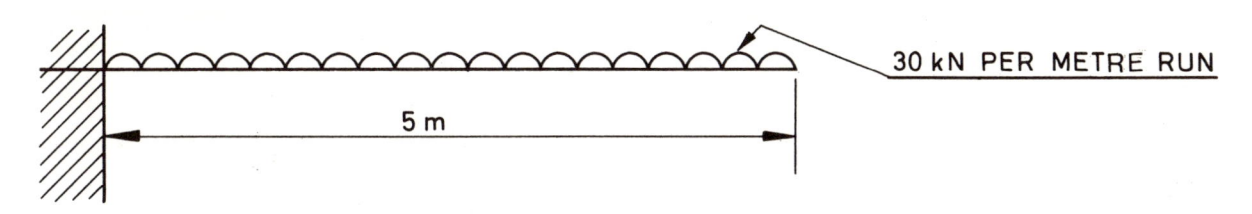

At the free end the shearing force is zero. At 1 metre to the left from the free end the load is 30kN, therefore the shearing force at this point is 30kN. At 2 metres from the right the load is 60kN (2 x 30), therefore the shearing force is 60kN. At 3 metres from the right the shearing force is 90kN, at 4 metres the shearing force is 120kN, and at the wall the shearing force is 150kN. At the wall there is an opposite reaction upwards, so that a vertical line drawn up to the base closes the diagram. Using a scale of 20mm=1m length of beam, and 1mm=3kN points are obtained for the shearing force at 1 metre intervals, and on joining these the shearing force diagram is obtained.

The bending moment at the free end is zero. At 1 metre from the free end the total load is 30 x 1=30kN, and its centre of gravity is 0.5m from the end. The bending moment at 1m from the right is therefore 30 x 0.5=15kNm. At a point 2 metres from the right the load up to this point is 30 x 2=60kN, and the centre of gravity of this load is 1m from the end, hence the bending moment is 60 x 1 = 60kNm. At 3 metres from the free end the load to this point is 30 x 3=90kN, and its C.G. is 1.5m from the end, hence the bending moment is 90 x 1.5=135kNm. At 4 metres the bending moment is 4 x 30 x 2=240kNm, and finally at the fixed end the bending moment is 5 x 30 x 2.5=375kNm. These bending moments are now plotted at their respective distances, and the points joined by a smooth curve as shown. The scale used in the bending moment diagram is 1 mm=10kNm.

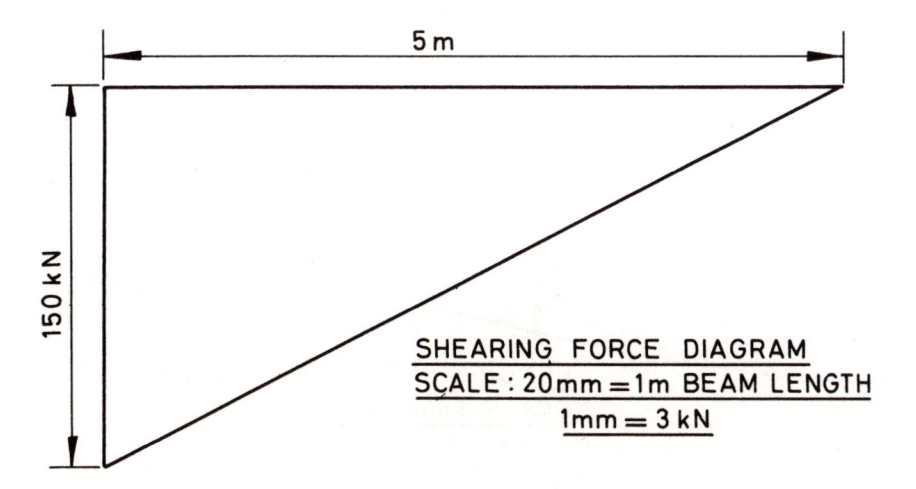

SHEARING FORCE DIAGRAM
SCALE: 20mm=1m BEAM LENGTH
1mm = 3 kN

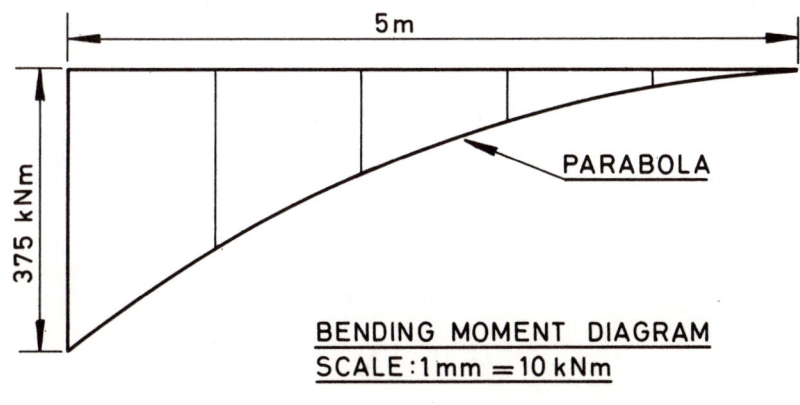

BENDING MOMENT DIAGRAM
SCALE: 1mm = 10 kNm

NOTE: IT IS USUAL TO DRAW S.F. AND B.M. DIAGRAMS ON A COMMON BASE. LIMITATION OF SPACE PREVENTS THIS IN THE ABOVE EXAMPLE.

Example

A beam 10 metres long is simply supported at each end, and carries a uniformly distributed load including the weight of the beam of 20kN per metre length. Draw the shearing force and bending moment diagrams.

Total load on beam=10 x 20=200kN and as it is uniformly distributed each reaction must carry 100kN. Using a scale of, say, 1mm=5kN, the shearing force diagram is constructed as shown below.

For the bending moment diagram which will be a parabolic curve, if say three positions are taken along the beam, at mid length, ¼ length and ¾ length, and the bending moments calculated for these, a fair curve through these points will represent the bending moment diagram: Obviously at each end the bending moment is zero, and is a maximum at the centre.

Bending moment at 2.5m from R_2
=(100 x 2.5)—(20 x 2.5 x 1.25)=250—62.5=187.5kNm.

Bending moment at mid length=(100 x 5)—(20 x 5 x 2.5)=500—250=250kNm.

Bending moment at 7.5m from R_2
=(100 x 7.5)—(20 x 7.5 x 3.75)=750—562.5=187.5kNm.

Scale of bending moment diagram 1mm=10kNm.

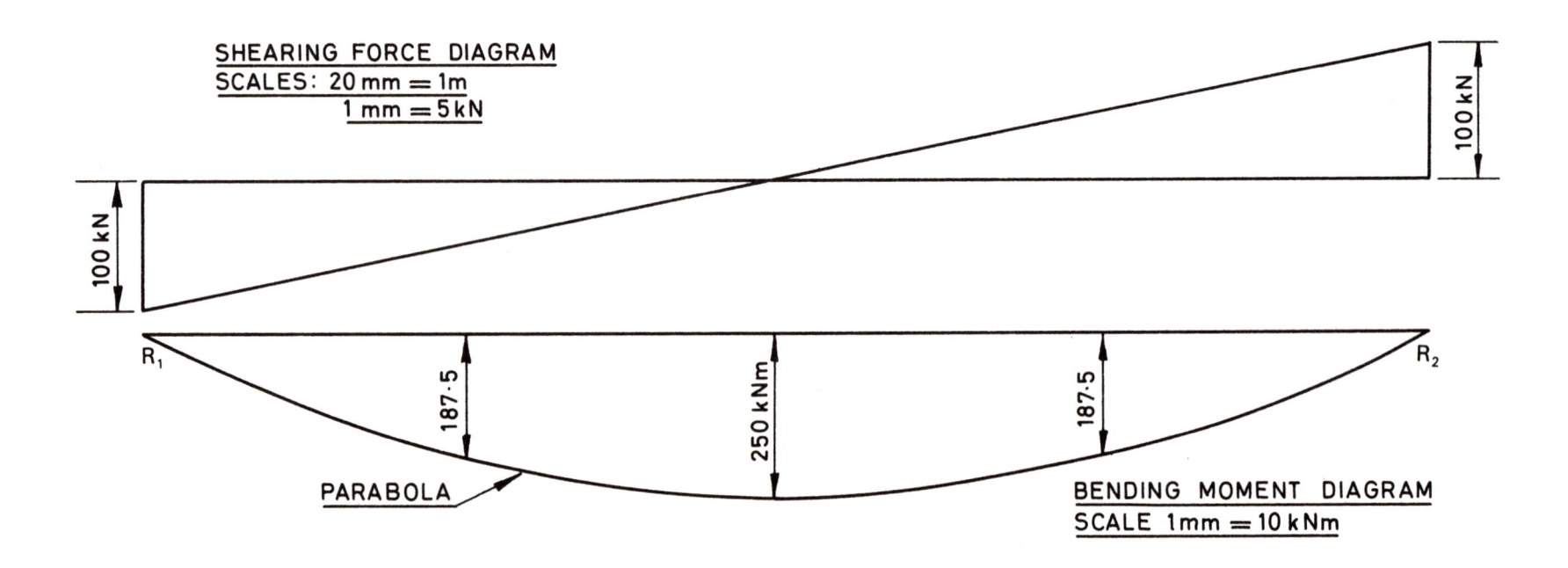

Example

A beam of 12 metres long is simply supported at each end and carries concentrated loads of 25 kN, 30 kN, and 50 kN at 4m, 6m, and 10m respectively from one end. Draw the shearing force and bending moment diagrams.

To find the reactions R_1 and R_2 moments are taken about either R_1 or R_2
Moments about R_1. Clockwise moments = anticlockwise moments

$$(25 \times 4)+(30 \times 6)+(50 \times 10) = R_2 \times 12$$
$$780 = 12 R_2$$
$$R_2 = 65 \text{ kN}$$
But $R_1 + R_2 = 25 + 30 + 50 = 105 \text{ kN}$
$$R_1 = 40 \text{ kN}$$

The shearing force diagram is constructed as shown using the following scales 15mm = 1m beam length. 1mm = 4 kN

To calculate the bending moment at each point of loading, it is only necessary to consider moments to either the right or the left of the point, so naturally one chooses the direction which has fewer forces.

Bending moment at 50 kN force (taking moments to the right)
$$= R_2 \times 2 = 65 \times 2 = 130 \text{ kNm}$$

Bending moment at midlength (taking moments to the right)
$$= (R_2 \times 6) - (50 \times 4) = (65 \times 6) - (50 \times 4) = 190 \text{ kNm}$$

Bending moment at 25 kN force (taking moments to the left)
$$= R_1 \times 4 = 40 \times 4 = 160 \text{ kNm}$$

The bending moment diagram is constructed as shown using a scale of 1 mm = 5 kNm

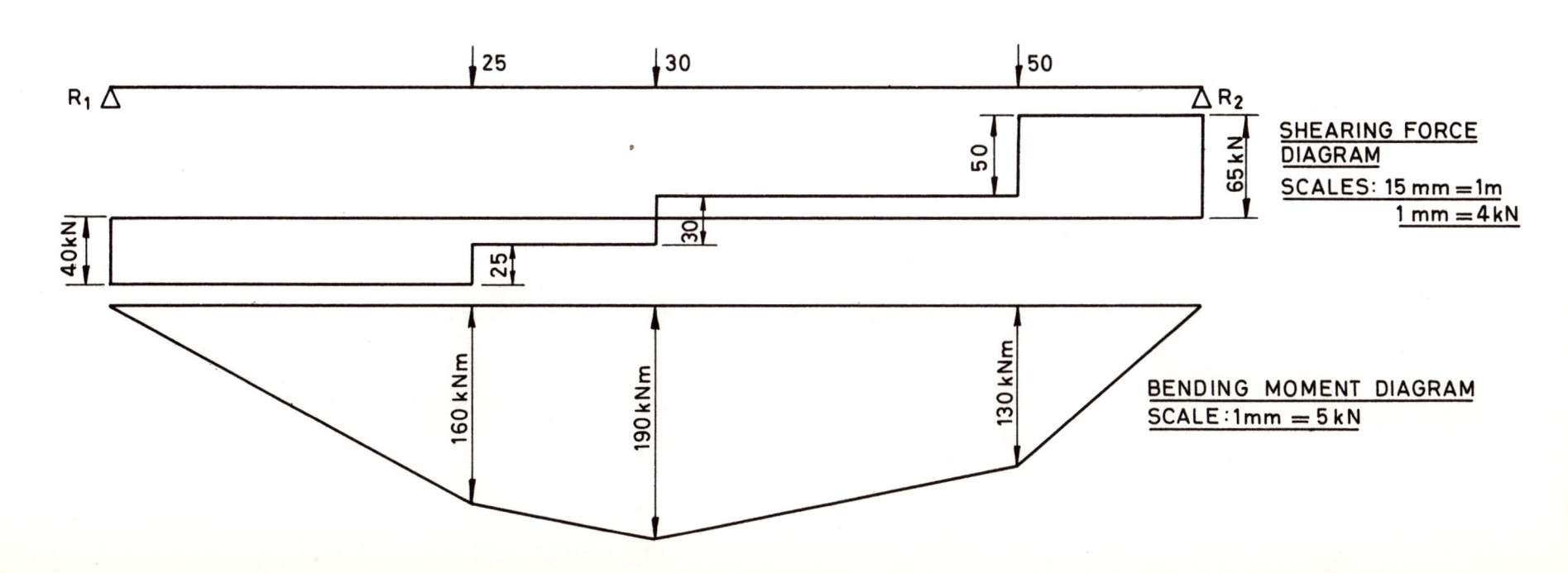

167

Example

A beam 16 metres long is simply supported at 3 metres from each end, and carries a uniformly distributed load of 12 kN per metre run. Draw the shearing force and bending moment diagrams, and measure off the position where the bending moment is zero.

Total load = 12 x 16 = 192 kN

Reaction at each support = $\frac{192}{2}$ = 96 kN

From the right the shearing force diagram starts at zero, and slopes at the rate of 12kN per metre for 3 metres to the reaction R_2. It then rises vertically a distance to represent R_2 = 96 kN. From this point there is a downward gradient of 12 kN per metre for 10 metres, then a vertical rise to represent R_1 = 96 kN. Finally to close the diagram there is a downward gradient of 12 kN per metre for the remaining 3m. Scale of shearing force diagram 10mm = 1m beam length. 5 kN = 1mm.

Bending moment at mid length = (96 x 5) — (12 x 8 x 4) = 96 kNm
Bending moment at reactions (in opposite direction to bending moment at mid length) = 12 x 3 x 1.5 = 54 kNm
These points are plotted to a scale of 5 kNm = 1mm and joined up by smooth parabolic curves.

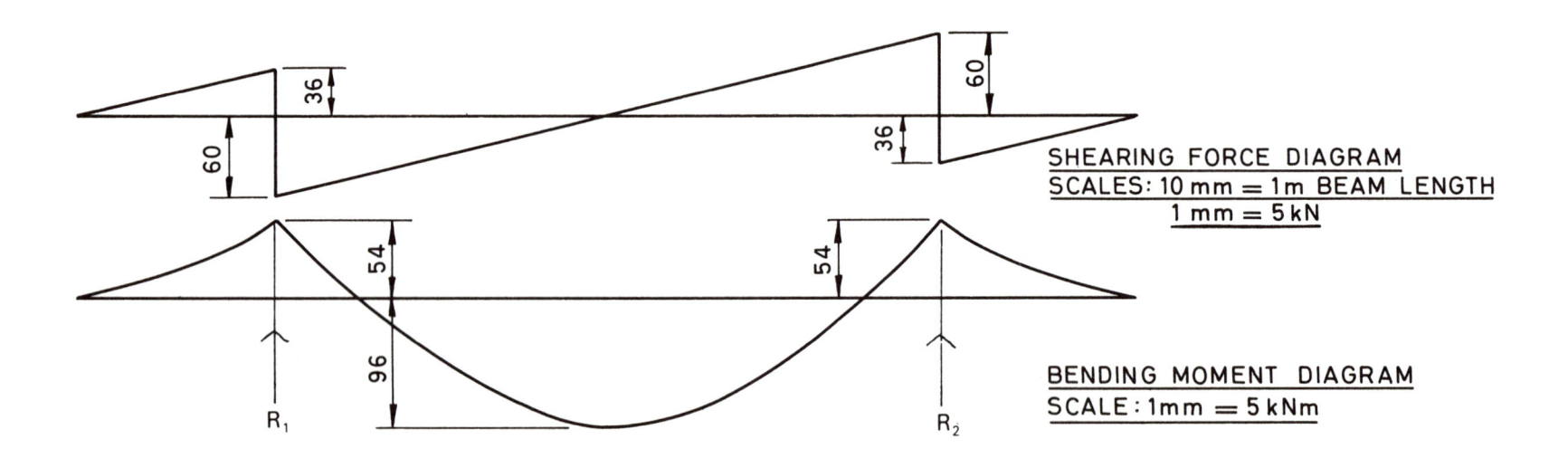

SHEARING FORCE DIAGRAM
SCALES: 10 mm = 1m BEAM LENGTH
1 mm = 5 kN

BENDING MOMENT DIAGRAM
SCALE: 1mm = 5 kNm

BENDING MOMENT IS ZERO AT 4.3m FROM EACH END

168

DIMENSIONS OF ISO METRIC
HEXAGON PRECISION BOLTS SCREWS AND NUTS

Nom Size	Pitch of thread Coarse	Pitch of thread Fine	Diameter of unthreaded shank max	Diameter of unthreaded shank min	Width across flats max	Width across flats min	Width across corners min	Depth of washer face	Radius under head max	Radius under head min	Height of head nom	Thickness of nut nom
M3	0.50	-	3.00	2.86	5.50	5.38	6.08	0.1	0.30	0.10	2.00	2.40
M4	0.70	-	4.00	3.82	7.00	6.85	7.74	0.1	0.35	0.20	2.80	3.20
M5	0.80	-	5.00	4.82	8.00	7.85	8.87	0.2	0.35	0.20	3.50	4.00
M6	1.00	-	6.00	5.82	10.00	9.78	11.05	0.3	0.40	0.25	4.00	5.00
M8	1.25	1.00	8.00	7.78	13.00	12.73	14.38	0.4	0.60	0.40	5.50	6.50
M10	1.50	1.25	10.00	9.78	17.00	16.73	18.90	0.4	0.60	0.40	7.00	8.00
M12	1.75	1.25	12.00	11.73	19.00	18.67	21.10	0.4	1.10	0.60	8.00	10.00
*M14	2.00	1.50	14.00	13.73	22.00	21.67	24.49	0.4	1.10	0.60	9.00	11.00
M16	2.00	1.50	16.00	15.73	24.00	23.67	26.75	0.4	1.10	0.60	10.00	13.00
M18	2.50	1.50	18.00	17.73	27.00	26.67	30.14	0.4	1.10	0.60	12.00	15.00
M20	2.50	1.50	20.00	19.67	30.00	29.67	33.53	0.4	1.20	0.80	13.00	16.00
*M22	2.50	1.50	22.00	21.67	32.00	31.61	35.72	0.4	1.20	0.80	14.00	18.00
M24	3.00	2.00	24.00	23.67	36.00	35.38	39.98	0.5	1.20	0.80	15.00	19.00
*M27	3.00	-	27.00	26.67	41.00	40.38	45.63	0.5	1.7	1.0	17.00	22.00
M30	3.50	-	30.00	29.67	46.00	45.38	51.28	0.5	1.7	1.0	19.00	24.00
*M33	3.50	-	33.00	32.61	50.00	49.38	55.80	0.5	1.7	1.0	21.00	26.00
M36	4.00	-	36.00	35.61	55.00	54.26	61.31	0.5	1.7	1.0	23.00	29.00
*M39	4.00	-	39.00	38.61	60.00	59.26	66.96	0.6	1.7	1.0	25.00	31.00
M42	4.50	-	42.00	41.61	65.00	64.26	72.61	0.6	1.8	1.2	26.00	34.00

*These are non-preferred diameters to be dispensed with wherever possible.
For further detail and large diameters refer to BS 3692-1967

INDEX

	Page
Abbreviations for General Engineering Terms	7
Air Inlet Valve	94, 95, 147, 156
Automatic Valve	52, 53, 126
Ballast Chest	42, 43, 121
Bilge Suction Strainer	24, 25, 112
Burner Carrier	56-59, 128, 129
Centrifugal Brake	62, 63, 131
Common Features — Representation of	9, 10, 11
Compressor Piston and Suction Valve	104, 105, 152
Connecting Rod and Bearings	60, 61, 130
Control Valve	30, 31, 115
Crane Hook	28, 29, 114
Crosshead and Guide Shoe	44, 45, 122
Curves — Connecting Rod Palm	16
Curves of Interpretation	15
Cylinder Relief Valve	26, 27, 113
Drawing Instruments Required	1
Ellipse — Construction of	12
Feed Check Valve	46, 47, 123
Flow Regulator	88, 89, 144
Forces — Parallelogram of	157
— Polygon of	160
— Reciprocating Engine Mechanism	161
— Representation by Vectors	157
— Triangle of	158

	Page
Fuel Control Lever	22, 23, 111
Fuel Valve	90, 91, 145
Full Bore Safety Valve	86, 87, 143
Gauge Glass (Plate Type)	106, 107, 153
Gear Pump	48, 49, 124
Helix — Construction of	14
Hexagonal Bolts and Nuts — Table of Dimensions	169
Hexagonal — Approximate Proportions	8
High Lift Safety Valve	84, 85, 142
Hydraulic Steering Fear	102, 103, 151
Involute — Construction of	13
Lines — Projection, Dimension and Leaders	6
— Types	3
Machined Block	20, 21, 110
Machining Fixture	20, 21, 110
Main Gear Wheel	100, 101, 150
Mechanical Lubricator	108, 109, 154
Michell Thrust Block	98, 99, 149
Oil Strainer	36, 37, 118

	Page
Parallel Slide Stop Valve	38, 39, 119
Pedestal Bearing	78, 79, 139
Piston (4 stroke)	70, 71, 135
Piston (Upper and Rod)	72, 73, 136
Piston Type Stop Valve	40, 41, 120
Projection — First Angle	4, 155
Projection — Third Angle	155, 156
Quick Closing Valve	64, 65, 132
Reducing Valve	68, 69, 134
Rudder Carrier Bearing	66, 67, 133
Scales — Recommended	6
Sealed Ball Joint	32, 33, 116
Shearing Force and Bending Moment Diagrams	163-168
Spur Gearing	17, 18
Starting Air Valve	54, 55, 127
Starting Air Pilot Valve	50, 51, 125
Stern Tube and Tailshaft	96, 97, 148
Telemotor Receiver	74, 75, 137
Tunnel Bearing	92, 93, 146
Turbine Flexible Coupling	80-83, 140, 141
Universal Coupling	34, 35, 117
Valve Actuator	76, 77, 138
Velocity Diagrams	162
Views — Sectional and Half Sectional	5